Handshake Between Chinese and Portuguese Design

中葡设计握手

家具篇
Furniture

陈静勇 等著
Chen Jingyong et al

中国建筑工业出版社

图书在版编目（CIP）数据

中葡设计握手 家具篇/陈静勇等著. --北京：中国建筑工业出版社，2018.10
ISBN 978-7-112-22388-6

I. ①中… II. ①陈… III. ①建筑设计-汉、英 ②家具-设计-汉、英 IV.①TU2 ②TS664.01
中国版本图书馆CIP数据核字（2018）第137675号

责任编辑：唐旭　吴佳
责任校对：芦欣甜

中葡设计握手　家具篇
陈静勇 等著

*
中国建筑工业出版社出版、发行（北京海淀三里河路9号）
各地新华书店、建筑书店经销
北京富诚彩色印刷有限公司
*
开本：880×1230毫米 1/16　印张：15¼　插页：1　字数：543千字
2018年10月第一版　2018年10月第一次印刷
定价：**208.00** 元
ISBN 978-7-112-22388-6
（ 32084 ）

内容提要

《中葡设计握手 家具篇》是一部对2015年中葡文化交流项目——第二期“中国青年设计师驻场四季计划”（简称“驻场计划”）和2016北京国际设计周“中葡设计握手·家具篇——从北京到帕雷德斯”展览暨中外交流会的纪实蓝本，同时又是对北京建筑大学在“传统家具制作技艺”类国家级非物质文化遗产研究领域复原完成的明式家具“标准器”代表作品和参加第二期“驻场计划”中葡双方青年设计师协同完成的创新家具作品、展陈设计等研究著作。

全书采用中英文对照，刊出大量精美的家具设计图样，突出设计文化元素。旨在呈现中葡两国设计文化交流新面貌，启发人们对文化传承与创新的思考。适宜作为设计艺术院校相关专业开展家具设计和对外合作与交流的借鉴参考。

Summary

"Handshake between Chinese and Portuguese Design. Furniture" It is a book about a cultural communication between China and Portugal, comprise the "Phase II young Chinese designers Design-in-Residence Programmes" (Design-in-Residence Programmes) and the exhibition of Handshake between Chinese and Portuguese Design. Furniture - From Beijing to Paredes in 2016 Beijing Design Week. The show displayed the Standard-style of Ming-style furniture duplicate by Beijing University of Civil Engineering and Architecture in the field of National Intangible Cultural Heritage in traditional furniture making skills and the innovative furniture work and exhibition design combine both young designers of China and Portugal. The book uses Chinese and English and posts a large number of excellent furniture picture. It is a purpose of showing the new look of communication of Chinese and Portuguese design so that it inspired the thinking of cultural heritage and innovation. This book appropriate to furniture design and foreign exchange in the school of design and arts to reference.

Editorial Committee 编委会

展览组织委员会

主办：
北京市人民政府外事办公室
北京建筑大学

承办：
中华世纪坛世界艺术中心
葡中文化协会
北京歌华文化发展集团

协办：
文化部恭王府管理中心
北京市非物质文化遗产保护中心
中国建筑学会室内设计分会
北京工业设计促进中心

展览设计：
北京歌华设计有限公司
北京建筑大学

Exhibition Committee

Host:
Foreign Affairs Office of the People's Government of Beijing Municipality
Beijing University of Civil Engineering and Architecture

Undertake:
China Millennium Monument · World Art Center
ACLC - Associacao Cultural Luso Chinesa
Beijing Gehua Cultural Development Group

Co-organizing:
The Museum of Prince Kung's Palace
Beijing Intangible Cultural Heritage Protection Center
China Institute of Interior Design
Beijing Industrial Design Center

Exibition Design:
Beijing Gehua Design Co,. Ltd
Beijing University of Civil Engineering and Architecture

序

保护与传承好世界非物质文化遗产，是保护好人类文化遗产的重要组成部分。2003年10月联合国教科文组织第32届大会通过《保护非物质文化遗产公约》，2006年4月生效，旨在保护以传统、口头表述、节庆礼仪、手工技能、音乐、舞蹈等为代表的非物质文化遗产。2012年2月《中华人民共和国非物质文化遗产法》颁布，对继承和弘扬中华民族优秀传统文化，促进社会主义精神文明建设，加强非物质文化遗产保护、保存工作，起到了法律保障作用。

中华文明和灿烂文化具有五千年悠久历史，中国也是世界非物质文化遗产大国，具有三千年建城史、八百年建都史的首都北京更是中华文明的金名片，《北京城市总体规划（2016年-2035年）》表述为“北京是见证历史沧桑变迁的千年古都，也是不断展现国家发展新面貌的现代化城市，更是东西方文明相遇和交融的国际化大都市。北京历史文化遗产是中华文明源远流长的伟大见证，是北京建设世界文化名城的根基”。

北京建筑大学办学历史悠久，学术沉积深厚，源于1907年清政府成立的京师初等工业学堂，当时就有木工科、金工科，我校始终坚守“北京味十足”“建筑味十足”的办学定位和特色优势。作为北京市与住建部共建高校、北京地区唯一的建筑类高校、“国家建筑遗产保护研究和人才培养基地”，我们正在努力建设国内一流、国际知名、具有鲜明建筑特色的高水平、开放式、创新型大学。我校开展的“传统建筑营造技艺”“传统家具制作技艺”等中国非物质文化遗产传统技艺类保护与传承研究项目和人才培养具有历史文化积淀。我校同文化和旅游部恭王府博物馆共建有“中国非物质文化遗产研究院”，2016年获批文化部　教育部“中国非物质文化遗产传承人群研修研习普及培训计划项目”，持续开展了传统建筑营造技艺、传统家具制作技艺类别项目的实施工作。我校设计艺术研究院陈静勇教授主持的传统家具构法与制作技艺研究所，多年来持续开展了传统家具制作技艺保护与传承研究，2016年主持完成恭王府丙申·寒露“国家级非物质文化遗产：苏州造物——明式家具（苏作）制作技艺精品展”的策展设计，编著出版研究专辑。

2015年6月-7月，以我校为主的设计艺术领域的青年教师和学生赴葡萄牙帕雷德斯市参加了由北京市人民政府外事办公室主办、北京歌华文化发展集团和中葡文化创意合作平台共同承办的第二期“中国青年设计师驻场四季计划”（以下简称“驻场计划”），并取得丰硕的国际合作与交流成果。这是北京市积极践行“文化走出去”、鼓励传播优秀传统文化和创意设计活动与市场的同步、着力打造“中外技艺凝聚中国设计、现代设计诠释传统文化”的品牌活动。

2016年9月9日，由北京市人民政府外事办公室和北京建筑大学联合主办，歌华设计有限公司承办的2016北京国际设计周“中葡设计握手·家具篇——从北京到帕雷德斯”展览暨中外交流会在北京成功举办，展出陈静勇教授主持完成的明式家具“标准器”复原研究作品和参加第二期“驻场计划”中葡双方青年设计师协同完成的创新家具作品，集中诠释出中萄两国设计交流文化内涵。陈静勇教授所著的这部《中葡设计握手　家具篇》是“中国青年设计师驻场四季计划”和北京国际设计周活动的在学术层面上专题学术延伸，更是北京高校在设计文化对外合作与交流上实现“握手”的代表性案例，值得宣传和推广。

2017年9月，我校成为北京市首批“一带一路”国家人才培养基地。10月10日，我校牵头成立“一带一路”建筑类大学国际联盟，首批已有俄罗斯、波兰、法国、美国、英国、亚美尼亚、保加利亚、捷克、韩国、马来西亚、希腊、尼泊尔、以色列等19个国家的44所大学加入联盟。该联盟旨在发挥特色优势、推进资源共享、加强协同创新、促进人才培养，服务“一带一路”沿线及欧亚地区的发展建设，标志着“一带一路”沿线建筑类大学的合作已经迈出了坚实的第一步，我们将进一步深化务实合作，推动互利共赢，创新合作交流机制，不断培养和造就“古都北京的保护者、宜居北京的营造者、现代北京的管理者、未来北京的设计者、创新北京的实践者”，为推进为北京的，为中国的，为世界的建筑文化遗产保护、非物质文化遗产传承、世界宜居城乡发展建设做出更大贡献。

二〇一八年五月二十一日（戊戌·小满）
“一带一路”建筑类大学国际联盟主席
北京建筑大学校长
张爱林

Preface

It is a significant part of the protection of human Cultural heritage which is concentrate on protection and heritage in World Intangible Cultural Heritage. October 2003, The Convention of Protect the Intangible Culture Heritage was published by UNESCO general assembly in 32nd, and it became active in April 2006. It protects the Intangible Culture Heritage which is represented by the oral presentation, festival etiquette, manual skills, music and dance. February 2012, Law of Intangible Cultural Heritage of People's Republic of China was published. It works for inheriting and popular the excellent traditional culture of Chinese, promoting the construction of socialist spiritual civilisation, strengthening the protection and preservation of intangible cultural heritage by law protection.

The splendid culture of Chinese have 5000 years of history, and also China is a beautiful country which has many World Intangible Cultural Heritage. Beijing, the capital of China, is a golden simple of China which has 3000 years history of urban construction and 800 years history of capital construction. General Planning for Beijing Municipality (2016-2035) said, "Beijing is the ancient capital of the Millennium in which witness the history of vicissitudes of life, also it is a modern city that shows the new development of China, and it is an international city that meets and blends the eastern and western. The history cultural heritage of Beijing is a great witness that experienced the Chinese civilisation, and it is the base of Beijing in contributing the world-famous cultural city. "

Beijing University of Civil Engineering and Architecture was from the Imperial Primary industry school which established by Qing dynasty government in 1907, and it has a long history of education and a deep academic deposition. At that moment, BUCEA had majors of woodworking and metalworking. BUCEA always insist on the features advantages of "Beijing-style" and "Architecture-style" school goals. As the school which is the only Architectural University in Beijing cooperate with Ministry of Housing and Urban-Rural Development of the People's Republic of China. Also, it is a base of talent training, protection and research of national architectural heritage. BUCEA is working hard to build a high-level, open, and innovative university with excellent, internationally-renowned, distinctive architectural features. Also, it has some research program of protection and inherent in traditional making skills of Chinese Intangible Cultural Heritage which are traditional building construction skills and traditional furniture making skills. BUCEA cooperate with the Museum of Prince Kung's Palace of Ministry of Culture and Tourism to set up the Institute of Chinese Intangible Culture Heritage. In 2016, it was approved to carry out the project named Study popularisation and training program of Chinese Intangible Culture Heritage with Inheritors that continuing work on traditional building construction skills and traditional furniture making skills. The Academy of Design and Arts Professor Chen Jingyong host the Institute of Traditional Furniture Structure Methodologies and Protection Techniques Research that keep working on protection and inheritance of traditional furniture making skills for many years. In 2016, the institute finished the exhibition design and published the research of National Intangible Heritage, BingShen year·Cold Dew, Su Zhou Zao Wu -

Traditional Su-style Furniture Making Crafts Boutique Exhibition in the Museum of Prince Kung's Palace.

Between June and July in 2015, BUCEA young teachers and students who are in the field of design and arts participated in the Design-in-Resident programs which mainly organised by Foreign Affairs office of the People's Government of Beijing Municipality, and undertook by Beijing Gehua Culture Development Group and ACLC-Associacao Cultural Luso Chinese. Moreover, it has a fruitful achievement in international cooperation and exchanges. It is a brand activity of Culture goes out by Beijing which is Chinese design combine with Chinese and foreign techniques, and modern design explains traditional culture made for encouraging spread unique traditional culture and creative design market.

September 9th, 2016, the exhibition and communication with Foreigners which is Handshake between Chinese and Portuguese Design. Furniture - From Beijing to Paredes was host by Foreign Affairs office of the People's Government of Beijing Municipality and Beijing University of Civil Engineering and Architecture, undertaking by Beijing Gehua Culture Development Group in 2016 Beijing Design Week. This exhibition showed the Standard Ming-style furniture which recovered by Professor Chen Jingyong and his group and the new furniture design by the cooperation of both Chinese and Portuguese designer who took part in the Phase || Design-in-Resident. It epitomised the meaning of the exchange of Chinese and Portuguese design culture. This book is an extent of the topic study of Beijing Design Week and program of Chinese designer. Moreover, it is a representative case in design handshake between the University of Beijing and the foreign country, also it worth to publicity and promotion.

In September 2017, our university became the first "The Belt and Road" national talent training base in Beijing. On October 10th, our university took the lead in establishing "The Belt and Road" architectural university international alliance. The first batch of countries has included Russia, Poland, France, the United States, the United Kingdom, Armenia, Bulgaria, the Czech Republic, South Korea, Malaysia, Greece, Nepal and Israel, 44 universities in each country joined the alliance. The alliance aims to give full attention to its unique advantages, promote resource sharing, strengthen collaborative innovation, promote talent cultivation, serve to the development of "The Belt and Road" and the Eurasian region, marking the cooperation of architectural universities along "The Belt and Road" initiative has taken a solid first step. We will further pragmatic cooperation, promote mutual benefit and win-win results, innovate cooperation and exchange principles, and continue to nurture and cultivate "the protectors of the ancient capital, the creators of livable Beijing, the managers of modern Beijing, the designers of future Beijing and the practitioners of innovative Beijing" Wave made more significant contributions to the advancement for Beijing, for China, the protection of the world's architectural cultural heritage, the inheritance of intangible cultural heritage, and the development of the world's livable urban and rural areas.

2018.05.21(WuXu Year · Grain buds)
Chairman of "The Belt and Road" Architectural University International Alliance
President of Beijing University of Civil Engineering and Architecture
Zhang Ailin

前言

1 依托一个中外文化交流平台

“文化是一个国家、一个民族的灵魂。文化兴国运兴，文化强民族强。没有高度的文化自信，没有文化的繁荣兴盛，就没有中华民族伟大复兴。”

为贯彻《国务院关于推进文化创意和设计服务与相关产业融合发展的若干意见》精神，实施《北京市推进文化创意和设计服务与相关产业融合发展行动计划（2015-2020年）》，积极践行“文化走出去”，鼓励传播优秀传统文化和创意设计活动与市场的同步，着力打造“中外技艺凝聚中国设计、现代设计诠释传统文化”的品牌活动。“中国青年设计师驻场计划”项目正是以此为目标和宗旨的对外文化交流活动。

“中国青年设计师驻场四季计划”（以下简称“驻场计划”）由北京市人民政府外事办公室主办，北京歌华文化发展集团和中葡文化创意合作平台共同承办。该项目旨在鼓励设计活动与工业生产的同步、鼓励原创产品和内容的创作、分享和传播，同时兼顾社会责任和可持续发展。

“驻场计划”源自由欧盟支持的“设计师居住项目”，该项目系欧盟主导的最具创新性和激励性的区域发展项目，在2012年获得了由欧洲委员会颁发的“RegioStars”奖项。在之前举办过的“设计师居住项目”中，来自西班牙、英国、智利、印度、新西兰、南非等多个国家的设计师参与其中，并取得了良好的反响。

北京作为该项目在华首个合作伙伴，2015年启动“驻场计划”，举办了中国青年设计师赴葡萄牙开展驻场访学，每期时长30天，旨在挖掘中国具有潜力的设计师与葡方合作，共同开发设计作品，并有机会与当地企业合作，为其提供设计服务，并由这些企业将优秀设计成果转化成产品进行批量生产，并在欧洲和世界各地进行销售。

2 参加一个家具设计专题项目

第二期“驻场计划”于2015年6-7月成功举行。北京市人民政府外事办公室组织了来自以北京建筑大学为主院校的10名设计艺术领域的青年教师和学生赴葡萄牙帕雷德斯市参与该项目，项目主题为家具设计。在葡期间，协同完成了设计作品，与当地设计艺术院校就设计人才培养进行了深入探讨。

中国设计师参与“驻场计划”，与葡方设计师交流设计理念，协同完成设计作品，这在中葡设计艺术交流领域尚属首次，葡萄牙当地媒体对此给予了高度关注。包括《葡萄牙商业周刊》《葡萄牙经济日报》在内的10余家葡萄牙媒体对此进行了详尽的报道，葡萄牙TVI电视台还就此制作了专题节目，反映驻场青年设计师团队在葡工作和生活情况，取得良好反响。

葡方对中方青年设计师的到来高度重视。其中第二期“驻场计划”中方设计人员于当地时间6月29日上午受到了葡萄牙时任总统卡瓦科·席尔瓦（Cavaco Silva）的亲切接见。席尔瓦总统对北京市代表团的到来表示热烈地欢迎；他表示自己曾去过北京和上海，对中国有着深厚的感情；询问了参加“驻场计划”师生的基本情况，并表示愿意进一步推进中葡间的文化交流。

中方青年设计师将北京建筑大学“传统家具构法与制作技艺（陈静勇教授）工作室”与文化部恭王府中华传统技艺研究与保护中心联合编著出版的《中华传统技艺3. 2014小暑卷：明式家具传统制作技艺学术研讨会暨明氏十六品高仿作品展特辑》一书作为礼物送给席尔瓦总统和帕雷德斯市市长塞尔索·费雷拉（Colso Ferreira）。席尔瓦总统与费雷拉市长分别做了签名回赠。在葡期间，北京建筑大学还与葡萄牙波尔图大学签署了合作办学协议，开启了本校与葡萄牙高校联合人才培养和国际交流的新阶段。

3 借助一个国际设计周品牌

每年一届的北京国际设计周由中华人民共和国文化部与北京市人民政府共同主办。北京国际设计周秉承“设计之都、智慧城市”的理念，以“设计+”为主题，分布于京津冀地区相关文化创意场所。设计周还重点关注设计对推动国家经济转型，优秀传统文化传承创新，构建高精尖产业结构，以及产业结构转型、优化、升级等方面的作用，已成为北京具有国际影响力和可持续发展的创意设计活动，是北京建设全国文化中心、科技创新中心、国际交往中心的重要抓手，也是北京疏解非首都功能、构建高精尖产业结构的重要支撑。

由北京市人民政府外事办公室和北京建筑大学联合主办，北京歌华文化发展集团和葡中文化协会联合承办的2016北京国际设计周“中葡设计握手·家具篇——从北京到帕雷德斯”展览于2016年9月9日在北京歌华大厦13层DSC展厅开幕，并举办了中外交流会。

莅临展览开幕式暨中外交流会的主要领导和嘉宾有：北京市人民政府外事办公室副主任高志勇、葡萄牙驻华大使若热·托雷斯·佩雷拉（Jorge Torres Pereira）、北京市国有文化资产监督管理办公室主任助理郑志宇、北京建筑大学校长张爱林、副校长汪苏、北京工业设计促进中心主任陈东亮、文化部恭王府中华传统技艺研究与保护中心主任孙冬宁、北京歌华文化发展集团总经理李丹阳、北京市朝阳区实验小学书记王存友。参加开幕式还有2015年赴葡萄牙帕雷德斯市圆满完成中葡文化交流第二期“驻场计划”专题项目的北京建筑大学师生设计师与北京市朝阳区实验小学教师设计师代表、中葡文化创意合作平台代表、北京建筑大学师生代表、中外媒体代表等。

展览陈列从北京歌华大厦一层大堂开始，巨型展览形象识别招贴悬挂在大堂上空，点燃了北京国际设计周热烈、浓厚的“设计节日”气氛。展厅中用中葡两国世界文化遗产做视觉引导，以第二期“驻场计划”工作过程为线索，通过实物、图片、文字、灯光等展陈设计手段，集中诠释出中葡两国设计交流内涵。

4 收获一个学术探索成果

“中葡设计握手·家具篇——从北京到帕雷德斯”展览，旨在促进国际合作与交流，依托北京国际设计周，呈现中葡两国设计文化交流新面貌，启发人们对文化传承与创新的思考。策展设计理念在于，探索服务北京“四个中心”尤其是文化中心建设、国际交往中心的设计艺术学科方向和突出学科特色，展示北京建筑大学在“传统家具制作技艺”类国家级非物质文化遗产研究领域复原的8件明式家具“标准器”代表作品和参加第二期“驻场计划”中葡双方青年设计师协同完成的10件（套）创新家具作品。

本书的编著出版，更是“中国青年设计师驻场四季计划”和北京国际设计周活动的在学术层面的专题延伸。内容既有纪实写照，又有研究阐释，是研究和传播家具设计文化和创意交流成果的代表性案例。

戊戌新春，愿通过本书，记录下中葡双方青年设计师“设计握手”的成果，也把北京建筑大学在设计艺术学科领域上的学术思考带到“一带一路”建筑类高校联盟，进一步密切国际合作与交流，把中国文化和和平友谊带给全世界。

二〇一八年二月四日（戊戌·立春）
北京建筑大学设计艺术研究院
陈静勇

Forward

1 Depend on a Cultural Exchange platform between China and Foreign countries

"Culture is a country and nation's soul. Our country will thrive only if our culture thrives, and our nation will be strong only if our culture is strong. Without full confidence in our culture, without a rich and prosperous culture, the Chinese nation will not be able to rejuvenate itself."

"Implementing Action Program of the Beijing Municipality on Further Promoting the Integrated Development with Relevant Industries of Cultural Creativity and Design Services (2015-2020) "was actualized. It was to implement "Implementing the Several Opinions of the State Council on Further Promoting the Integrated Development of Relevant Industries of Cultural Creativity and Design Services" spirit and actively fulfill. Culture goes global.Encourage the dissemination of outstanding traditional culture and creative design activities and market synchronously, and strive to build brand activity which is Chinese design converging Chinese and foreign skills and traditional culture interpretation by modern design.

"Design-in-Residence Programs of Chinese Young Designers in Four Seasons" (Design-in-Residence Programs) host by Foreign Affairs Office of the People's Government of Beijing Municipality, and jointly undertook by Beijing Gehua Culture Development Group and Chinese and Portuguese Cultural Innovation Cooperation Platform. This Program proposes to encourage creating, sharing and spreading original production in design activity and industrial production, and also consider social responsibility and sustainable development.

"Design-in-Residence Programs" stem from "Designer residence project" which was supported by European Union.This project is a creative and incentive area development project and won the RegioStars Award which conferred by European Commission in 2012.It obtained a good repercussion in "Designer residence project" which hosted before. There were many designers participated in form different countries, such as Spain, England, Chile, India, New Zealand and South Africa.

Beijing launched "Design-in-Residence Programs" in 2015 because it was the first partner in China with this project.Chinese Young designers visit Portugal in 30 days. It proposes to dig out the most potential designer and give an opportunity to the local Portuguese enterprise to work together. Sale productions around Europe and the entire world by those enterprises which transform the excellent design work into batch production.

2 A Participation in a furniture design project

Phase II "Design-in-Residence Programs" succeed in June to July 2015. Foreign Affairs Office of the People's Government of Beijing Municipality organized ten young teachers and students who were in the field of design to participation in this furniture topic program in Paredes, Portugal. They had discussions about training for the young designer with some local design art colleges and finished the design works during the period in Portugal.

Chinese designers exchanged their ideas with Portuguese designers and finished the design works together during the program. It was the first time in the domain of design exchange between China and Portugal, and this program with high attention by local Portuguese press. This event gained detailed reports by more than 10 Portuguese press including Jornal de Negócios and Diário Económic. Moreover, Portuguese TV Station made a unique program to reflect the situation of the young designer group working and studying in Portugal, and it produced a good echo.

The Portuguese pay high attention on the young Chinese designer was coming. The Portuguese President Cavaco Silva met the Chinese designers in 29th June local time during the Phase II "Design-in-Residence Programs," and he gave a warm welcome to the delegation of Beijing. He said he had been to Beijing and Shanghai and had a profound feeling with China. Besides, he asked some base condition with teachers and students who were in this program and expressed it was willing to go a step further in culture communication between China and Portugal.

Chinese young designers gave a book which compiled by BUCEA professor studio of traditional furniture structure and production skills research and Chinese traditional skills research and protection center in the museum of Prince Kung's palace, named Chinese traditional skills 3. 2014 volume slight heat: traditional craft symposium of Ming style furniture and Ming shi sixteen high imitation works exhibition, as a present to President Cavaco Silva and Paredes mayor Colso Ferreira and they return by sign.Beijing University of Civil Engineering and Architecture and the University of Porto have reached the agreement on issues of education cooperation.It created the new stage in international communication and talent training with Portuguese universities.

3 Support by an international design week brand

Beijing design week host by Ministry of Culture of The People's Republic of China and The People's Government of Beijing Municipality. Beijing Design Week distributed in cultural and creative place within Beijing-Tianjin-Hebei region, theme "design+," follow the "Design city, Smart city" idea. Design week focus on the effect of national economic transformation by design impact, inheritance and innovation of excellent traditional culture, construction of the High-grand industrial structure and Industrial structure Transformation, Optimization and upgrading. Beijing Design week became a creative design activity which has international influence and sustainable development in Beijing. It is the most significant part of construct national culture center, science innovation center and international communication center in Beijing. Also, it is the dramatically support on dredge non-capital functions and Construction of High-grand Industrial structure in Beijing.

" Handshake between Chinese and Portuguese design. Furniture - from Beijing to Paredes" exhibition host by Foreign Affairs Office of the People's Government of Beijing Municipality and Beijing University of Civil Engineering and Architecture. Beijing Gehua Cultural Development Group and ACLC - ASSOCIAÇÃO CULTURAL LUSO - CHINESA jointly sponsored it. It undertook at DSC display hall, 13 floors, Beijing Gehua mansion in 9th September 2016. At the same time, communication between Chinese and foreigner held.

The principal leaders and guests attending the opening ceremony and communication: Foreign Affairs Office of the People's Government of Beijing Municipality Deputy Director Gao Zhiyong, Ambassador of Portugal to China Jorge Torres Pereira, State-owned Cultural Assets Supervision and Administration Office of the People's Government of Beijing Municipality Assistant Director Zheng Zhiyu, Beijing University of Civil Engineering and Architecture Headmaster Zhang Ailin, Beijing University of Civil Engineering and Architecture Vice President Wang Su, Beijing Industrial Design Center Director Chen Dongliang, Chinese Traditional Skills Research and Protection Center of The Museum of Prince Kung's Palace Director Sun Dongning, Beijing Gehua Cultural Development Group General Manager Li Danyang and Beijing Chaoyang Experimental Primary School Secretary Wang Cunyou. Also, participate in the opening were the teachers and students in BUCEA who took part in the Phase || "Design-in-Residence Programs." Beijing Chaoyang Experimental Primary School teacher delegates, Chinese and Portuguese Cultural Innovation Cooperation Platform delegates, teacher and student delegates of and Chinese and foreign media delegates and so on.

The exhibition started from the lobby at the first floor, Beijing Gehua mansion, huge placards were hung in the hall that lighted the enthusiastic and grumous "design festival" atmosphere during the Beijing Design Week. Exhibition used Chinese and Portuguese world cultural heritage to be visual guidance and utilized work process to be a clue. It Interpreted connotation about design exchange between Chinese and Portuguese through using production, picture, word, and light.

4 Achieve an academic exploration

"Handshake between Chinese and Portuguese design. Furniture - from Beijing to Paredes" exhibition proposes to promote the international cooperation and exchange. It depends on Beijing Design Week to illustrate the new look between Chinese and Portuguese design culture and inspire people thinking about culture inheritance and innovation. Exhibition design idea from exploration and service for Beijing "Four centers," especially Art &Design unique feature development in culture center and international communication center. The exhibition displayed 8 Ming-style Furniture which in the field of the traditional craft of furniture, State - level intangible cultural heritage, recovered by BUCEA, and 10 innovative furniture works created by both Chinese and Portuguese designers.

There was a particular topic which extended the academic level of Beijing design week activity and "Programs of Chinese Young Designers Design in Residence in Four Seasons." The content not only has documentary elements but also have Research interpretation. It is a representative case in the result of furniture culture innovation exchange of spreading and researching.

Write down the achievement of the handshake between Chinese and Portuguese designers in Wu Xu New Year.Hoping it can bring the academic thinking of BUCEA Art&Design subject into Belt and Road Architectural University Alliance and connect the international cooperation and communication more closely, Sending Chinese cultural, Peace and friendship to the entire world.

2018.02.04(WuXu Year · Spring begins)
Academic of design and arts of Beijing University of Civil Engineering and Architecture
Chen Jingyong

2015年6月29日，葡萄牙时任总统卡瓦科·席尔瓦（Cavaco Silva）在帕雷德斯创新孵化器亲切接见第二期"设计计划"北京代表团。
June 29, 2015, Portuguese President Cavaco Silva was in Paredes Innovation Incubator, and he cordially met with the Beijing delegation of "Phase II of Design in Residence Programs".

席尔瓦总统在北京建筑大学编著出版物上签字留念
President Silva signed on the publication edited by Beijing University of Civil Engineering and Architecture.

2016年9月9日，北京建筑大学张爱林校长出席2016北京国际设计周"中葡设计握手·家具篇——从北京到帕雷德斯"展览，并发表讲话。

On September 9th, 2016, Zhang Ailin the President of Beijing University of Civil Engineering and Architecture attended the Beijing Design Week 2016 "Handshake between Chinese and Portuguese Design. Furniture-From Beijing to Paredes" exhibition and delivered a speech.

中
Handshak
从 / 北
FROM
2015
立:
人民政府外事办公室
筑大学
立:
华设计有限公司
ffairs Office of Beijing
overnment
iversity of Civil Engin
chitecture
zer:
hua Design Compa

2016年9月9日，葡萄牙驻华大使若热·托雷斯·佩雷拉（Jorge Torres Pereira）参观"中葡设计握手·家具篇"展览，与北京建筑大学校长张爱林及师生们交流。
On September 9, 2016, Portuguese ambassador to China Jorge Torres Pereira visited the "Handshake between Chinese and Portuguese Design. Furniture" exhibition and communicated with the President of Beijing University of Civil Engineering and Architecture, Zhang Ailin, and the teachers and students.

葡萄牙驻华大使若热·托雷斯·佩雷拉在北京建筑大学编著出版物上签名留念
The Portuguese ambassador to China, Jorge Torres Pereira, signed on the publication edited by Beijing University of Civil Engineering and Architecture.

帕雷德斯市市长塞尔索·费雷拉（Celso Ferreira）亲切接见第二期“驻场计划”中方设计师
The Mayor of Paredes Celso Ferreira cordially met with the Chinese designers of "Phase II of Design-in-Residence Programs".

The CITY of CHAMPIONS
WELCOMES
YOU
California Missions

参加第二期“驻场计划”10位中方青年设计师
（左起：赵雄韬、郗鑫鑫、孙滨、邹乐、赵安路、闫卓远、沈翔宜、韩风、杨琳、朱宁克）

Participated in the "Phase II of Design-in-Residence Programs" 10 Chinese young designers
(From the left:Zhao Xiongtao, Xi Xinxin, Sun Bin, Zou Le, Zhao Anlu, Yan Zhuoyuan, Shen Xiangyi, Han Feng, Yang Lin, Zhu Ningke)

2016年9月9日，北京市人民政府外事办公室副主任高志勇，葡萄牙驻华大使若热·托雷斯·佩雷拉，北京市国有文化资产监督管理办公室主任助理郑志宇，北京建筑大学校长张爱林为2016北京国际设计周“中葡设计握手·家具篇”（从北京到帕雷德斯）展览开幕剪彩。

On September 9, 2016, Gao Zhiyong, deputy director of the Foreign Affairs Office of the Beijing Municipal People's Government, Portuguese Ambassador to China Jorge Torres Pereira, Assistant Director Zheng Zhiyu of the Beijing State-owned Assets Supervision and Administration Office, and Zhang Ailing, President of Beijing University of Architecture cut the ribbon for the Beijing Design Week 2016 "Handshake between Chinese and Portuguese Design. Furniture" (from Beijing to Paredes) Opening Ceremony.

手 家具篇
Furniture

ART 艺

RETHINK 悟

INTRODUCTION 简介

索引
Index

后记
Postscript

中葡家具设计演变脉络图
The evolvement of Chinese and Portuguese
furniture design

FIND

寻

席地而坐
Sit on the ground

在商、周两代的铜器里的禁、俎，反映了中国早期家具的雏形。家具以席、床、几、案、屏、箱等低矮型家具为主。
Jin and Zu which in the bronze of Shang and Zhou dynasty reflected the base of early Chinese furniture. Main furniture types were mat, bed, small table, screen, box and other low-type furniture.

过渡时期
Period of transition

以战国、秦汉、三国时期至晋、南北朝时期为代表是中国家具发展的过渡时期。该时期仍以席地而坐为主要起居方式。由于受到外来文化影响，胡床普及开来，也出现了如墩、椅、凳等高型家具。
The period of the Warring States, Qin dynasty, Han dynasty, Three Kingdoms, Jin dynasty and the Southern and Northern dynasties were the period of transition. This period still sat on the floor as the primary way of living. But the furniture had influenced by foreign culture, then Huchuang was more popular. There was some high-type furniture appeared. Such as stool and chair.

宋代是中国高型家具的大发展时期，将人们起居习惯由席地而坐转变为垂足而坐。宋代家具借鉴了木构建筑梁架结构，强调简洁秀丽、文雅轻巧、实用美观。
There was a great change that made a habit of Chinese people from sitting low-type furniture to high-type furniture in Song dynasty. Song dynasty furniture reference to the structure of wooden building beams, and it focuses on conciseness, legerity, practicability, and aesthetics.

垂足而坐
Pedal sitting

葡萄牙王国建立
The establishment of the Portuguese Kingdom

葡萄牙自1143年脱离西班牙成为独立王国。
Since 1143, Portugal became a country which independent from Spain.

明清时期中国家具的品类齐备，不拘泥于满足功能，同时造型上具有极高的艺术成就，有“结构严谨，线条简洁流畅，做工精湛，造型典雅隽秀，尺寸与比例科学合理”等艺术特色，因工具、工艺、材料等的成熟发展，中国传统家具技艺达到了世界家具艺术的巅峰。

Furniture design reached the peak of the Chinese traditional furniture art. Because of the mature development of tools, crafts, materials and so on, in Ming and Qing dynasties. In the period of that time, there are full of the classified of furniture and had extremely high artistic achievements in the shape which had strict structure, smooth and straightforward lines, exquisite craft, elegant appearance and reasonable size not only the practicability.

鼎盛时期
Heyday

清代中后期的“康乾盛世”，中西交流频繁，家具制作融合中外风格，装饰材料种类繁多，装饰手法采用雕刻、镶嵌、髹漆、彩绘等多种工艺手法相结合，呈现出雍容华贵、雕琢繁缛的风格。

Because the frequent exchanges between China and West, furniture absorbed both native and foreign styles, gorgeous and complicate had shown in a wide variety of decorative materials. And decorative techniques using carving, inlaying, lacquer painting, painting, etc. The period in the middle of Qing Dynasty called"Kang Qian flourishing age."

大航海时代的开启
The opening of Great Nautical Age

葡萄牙人开启大航海时代，建立了欧洲和亚洲之间的海上联系，将中国的书籍、瓷器、丝绸、绘画、纸张和家具等传到了葡萄牙。

The Portuguese opened the era of excellent navigation and set up maritime links between Europe and Asia. Then they spread Chinese books, porcelain, silk, painting, paper, furniture and other things to Portugal.

欧亚海上贸易
Maritime trade in Europe and Asia.

葡萄牙家具的一个黄金时代，内外贸易频繁，黄金、钻石等财富充足，促使欧洲艺术文化大繁荣。
It is a golden age of Portuguese furniture. Various domestic and foreign trade like gold, diamonds and other wealth, has prompted the prosperity of European art and culture.

宫廷家具影响民间，中国传统家具在家具品类、款式、用材以及结构、工艺等方面均发生了根本的变化，这些变化集中地反映在当时盛行的“海派家具”上，并初步形成专业分工的企业群。
China palace furniture influenced the folk. Chinese traditional furniture was changed in the furniture category, style, material and structure, technology and other aspects. These changes reflected in the prevailing "Shanghai furniture," and initially formed a professional division of enterprise group.

文化发展时期
The period of cultural development

文化交流时期
The period of cultural exchange

文化融合，科技融合
Cultural fusion,
technology integration

未来发展时期
The future
development period

中国家具携带自身悠久厚重的文化，结合最新科技，不断发展变化。
Chinese furniture carries its own long and thick culture, combined with the latest technology, and constantly develops and changes.

葡萄牙开始着重家具消费者的情感，建立服务全球的家具设计品牌。
Portugal began to focus on the emotions of the furniture consumers and set up global furniture design brands.

葡萄牙从设计、科学领域培养青年创意人才，教育机构和地方研究中心发挥作用。
Portugal develops training young creative talents from the fields of design and science, educational institutions and local research centers play a role.

TRAVEL

葡萄牙总统席尔瓦会见“中国青年设计师驻场四季计划”代表团

北京市人民政府外事办公室

应葡萄牙帕雷德斯市政厅邀请，北京市人民政府外事办公室主办、北京歌华美术公司承办并组织实施的第二期“中国青年设计师驻场四季计划”（以下简称“驻场计划”）代表团于当地时间2015年6月28日下午抵达葡萄牙进行访问交流。代表团一行受到葡方高度重视和当地媒体的高度关注。当地时间6月29日上午，葡萄牙总统席尔瓦在葡萄牙帕雷德斯创新孵化器会见了代表团一行。市政府外办、北京建筑大学、北京市朝阳实验小学的代表和参加第二期“驻场计划”的青年设计师共15人出席会见活动。

席尔瓦总统对北京市代表团的到来表示热烈地欢迎。他说，2015年3月，首期“驻场计划”在葡萄牙顺利举行。来自中国设计师们所展现出来的积极、乐观、认真、严谨和努力的态度给葡方留下了深刻印象。席尔瓦总统说，他曾访问过中国的北京和上海，对中国有很深的感情，愿意进一步推进中葡间的文化交流。

中方代表团团长、北京建筑大学副校长汪苏表示，“中国青年设计师驻场四季计划”为中葡两国，尤其是为北京市与葡萄牙帕雷德斯市文化创意和设计交流方面的合作提供了良好的平台。中方非常重视这个交流项目，派出了中国优秀青年设计师参加访学交流，中方感谢葡方为交流活动的顺利开展所做的努力。

“驻场计划”项目源于2014北京国际设计周葡萄牙家具设计展——“椅子上的艺术”取得了巨大成功。展览主办方葡萄牙帕雷德斯市希望以此为契机，进一步促进中葡两国间的文化交流。北京市市政府外办积极响应，促进双方建立中葡文化交流平台，开展“驻场计划”项目，为本市“文化走出去”提供有力支持。

下一步，北京市市政府外办将继续推动“驻场计划”项目后续发展，依托中葡文化交流平台，将该项目打造成本市将文化交流、文化贸易和外宣工作有机结合的品牌活动。同时，将积极发掘更多优质资源，努力推动涉外文创类项目落户北京，积极促进对外文化交流活动，提升北京市的国际化水平和城市形象。

Portuguese President Silva met with the delegation of the Design in Residence Programs of Chinese Young Designers in Four Seasons

Foreign Affairs Office of the People's Government of Beijing Municiplity

With the invitation of Portuguese Paredes' city hall, the delegation of the second plan named "Design-in-Residence Programs of Chinese Young Designers in Four Seasons" (Design-in-Residence Programs), arrived in Portugal on the afternoon of June 28th local time for a visit. The plan's sponsor was the Foreign Affairs Office of the People's Government of Beijing Municiplity, and Beijing Gehua art company played as the contractor responsible for organization and Implementation. This visit highly valued by Portugal and attracted the high attention of the local media. On the morning of June 29, 2015, local time, Portuguese President Silva met with the delegation at the innovation incubator in Paredes. The representatives of the foreign affairs office of municipal government, Beijing University of Civil Engineering and Architecture, Beijing Chaoyang experimental primary school, as well as the young designers, participated in the second "Design-in-Residence Programs" attended the meeting.

President Silva gave a warm welcome to the arrival of the Beijing delegation. He said that in March 2015, the first "Design-in-Residence Programmes" was held in Portugal. Chinese designers had shown positive, optimistic, earnest, rigorous and hard work attitude, which left a deep impression to Portugal. He also said he had visited China in Beijing and Shanghai. With deep feelings for China, he wished further promotion to cultural exchange between China and Portugal.

The Head of Chinese delegation Wang Su, who is also the vice-president of Beijing University of Civil Engineering and Architecture said, the young Chinese designers in the "Design-in-Residence Programmes" provided an excellent platform of the two countries in city culture creativity and cooperative design communication, especially for Beijing and Paredes. Being grateful for Portugal's efforts in the smooth development of the exchange activities, China attached much significance to this exchange program and sent outstanding young Chinese designers for exchanging.

The "Design-in-Residence Programs" came from the Portuguese furniture design exhibition in Beijing international design week 2014, which had achieved great success in the art of the chair". The organizer of the exhibition Paredes-a Portugal city-hoped to take this opportunity to promote further cultural exchange between China and Portugal. The foreign affairs office of municipal government responded positively. To promote the establishment of the cultural exchange platform between China and Portugal, the "field plan" was held, which provided strong support for the "cultural export" of Paredes.

Next, the foreign affairs office of municipal government would continue to promote the follow-up development of "Design-in-Residence Programs." Relying on the ACLC - ASSOCIAÇÃO CULTURAL LUSO - CHINESA, the project would go into a brand activity, which represents the organic combination of city cultural exchange, cultural trade, and international publicity. In the meantime, Paredes would actively explore more high-quality resources. The city would strive to promote the establishment of foreign culture creative projects in Beijing, and cultural exchanges with other countries, to enhance the international city level and image.

第二期“驻场计划”路线图

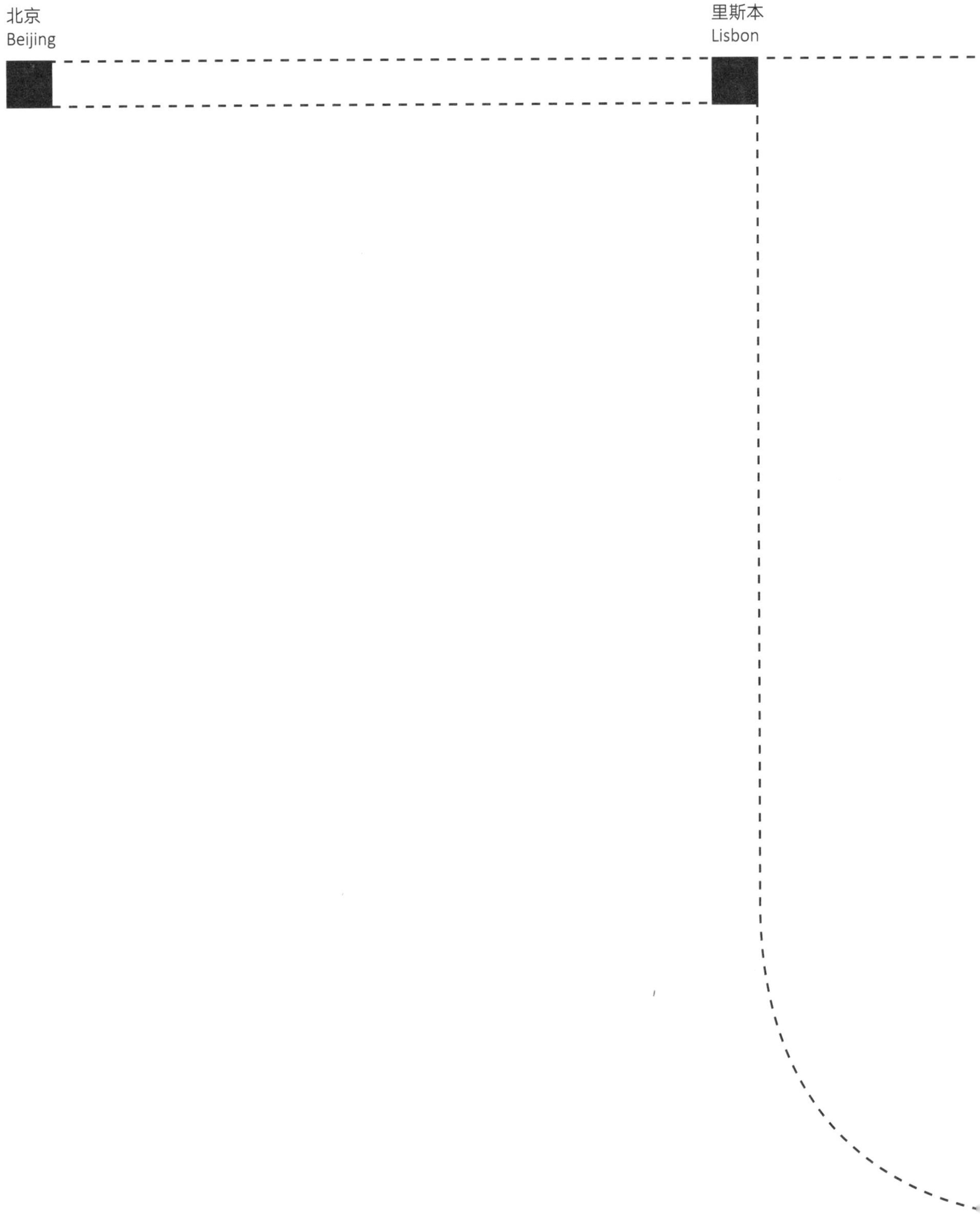

北京
Beijing

里斯本
Lisbon

Route of Phase II "Design-in-Residence Programs"

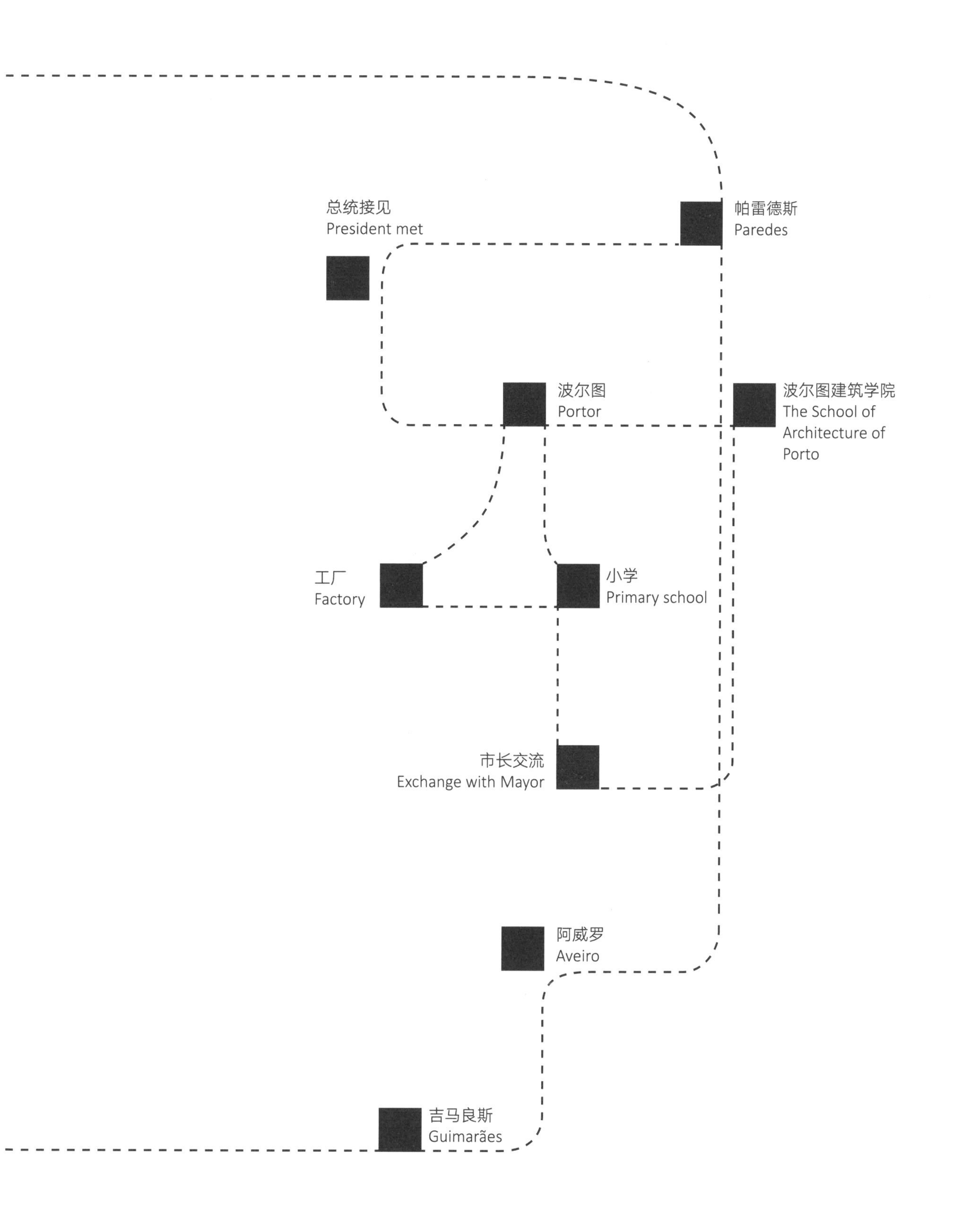

06/28 2015
帕雷德斯
Paredes

06/29
帕雷德斯
Paredes

06/30
波尔图
Porto

07/08
波尔图
Porto

07/09
波尔图
Porto

07/10
波尔图
Porto

07/11
阿威罗
Aveiro

07/12
阿威罗
Aveiro

07/13
波尔图
Porto

07/14
波尔图
Porto

07/15
帕雷德斯
Paredes

07/16
帕雷德斯
Paredes

07/17
帕雷德斯
Paredes

07/05
里斯本
Lisbon

07/06
帕雷德斯
Paredes

07/07
帕雷德斯
Paredes

07/01
帕雷德斯
Paredes

07/02
帕雷德斯
Paredes

07/03
帕雷德斯
Paredes

07/0
里斯本
Lisbon

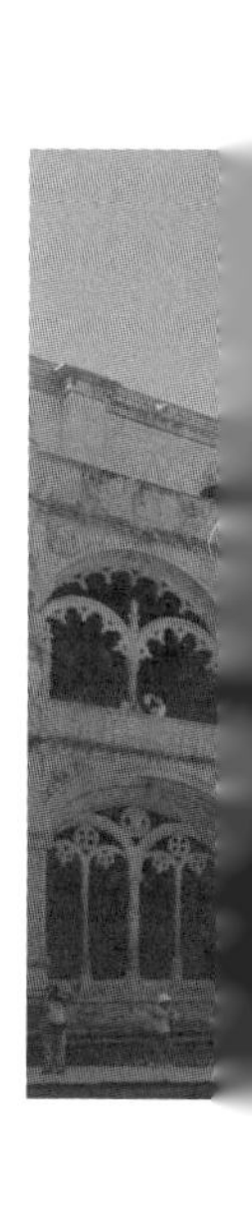

07/18
帕雷德斯
Paredes

07/19
波尔图
Porto

07/20
吉马良斯
Guimarães

07/21
吉马良斯
Guimarães

07/22
帕雷德斯
Paredes

07/23
帕雷德斯
Paredes

MAKE

造

四出头官帽椅
Armchair with four protruding ends

此椅原型为“明式四出头官帽椅”，是我国明式家具中椅子造型的一种典型款式。因其造型像古代官员的帽子而得名，上下无一丝附加装饰，结构简练之极，完全采用直线和弧线组合，线条曲直相同，方中带圆，充分体现了明式家具简洁明快的特点。

规格
长 x 宽 x 高　589 x 493 x 1127（mm）

复原者
北京建筑大学 工业设计111班
指导教师：陈静勇 郎世奇 尚华

The prototype of the chair is " official's hat armchair with four protruding ends of Ming-style." It is a very classic chair among the Ming-style furniture. The appearance of the chair is like official's hat in ancient times. It is concise and straightforward, with no decoration. The curve and straight lines of the chair fully express the simple style of Ming furniture.

Dimension
length x width x height 589 x 493 x 1127 (mm)

Maker
Beijing University of Civil Engineering and Architecture
Industiral Design Class of 111
Advisory teachers:Chen Jingyong, Lang Shiqi, Shang Hua

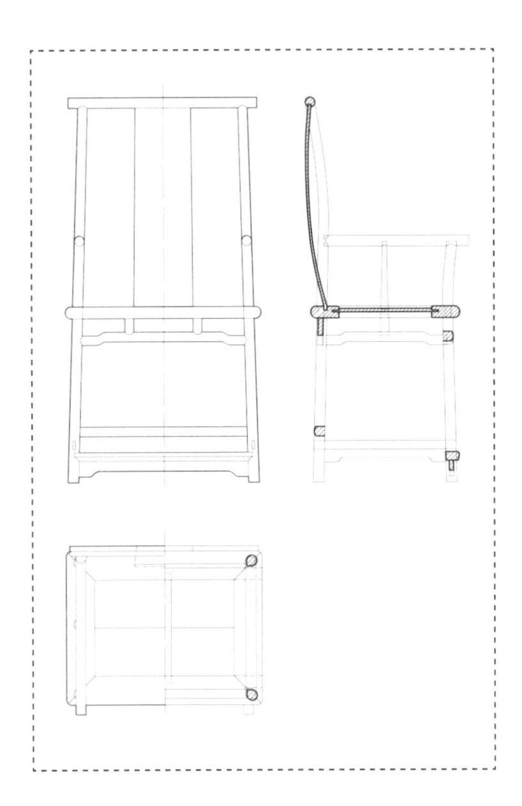

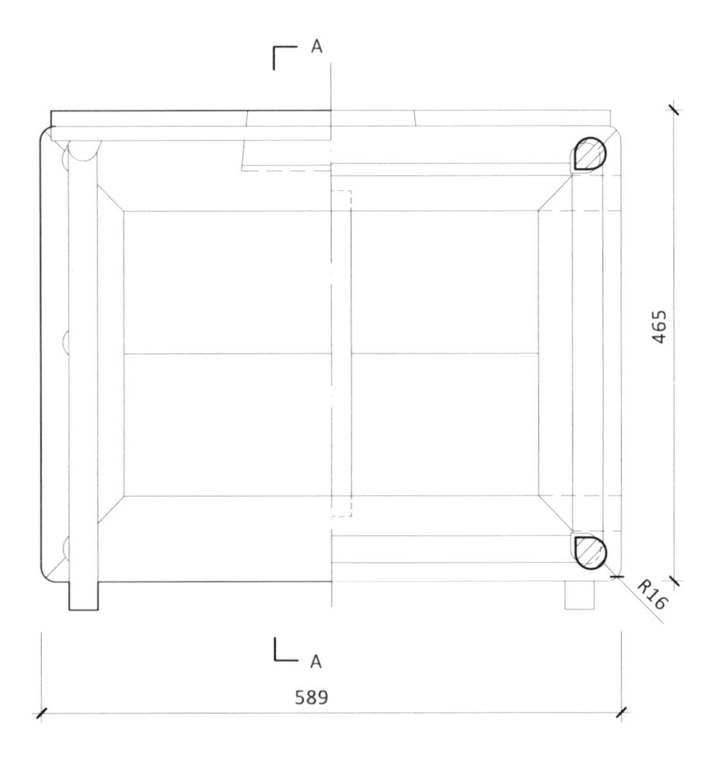

顶视图 · PLAN OF SEAT · B剖面图 · PLAN AT B ·

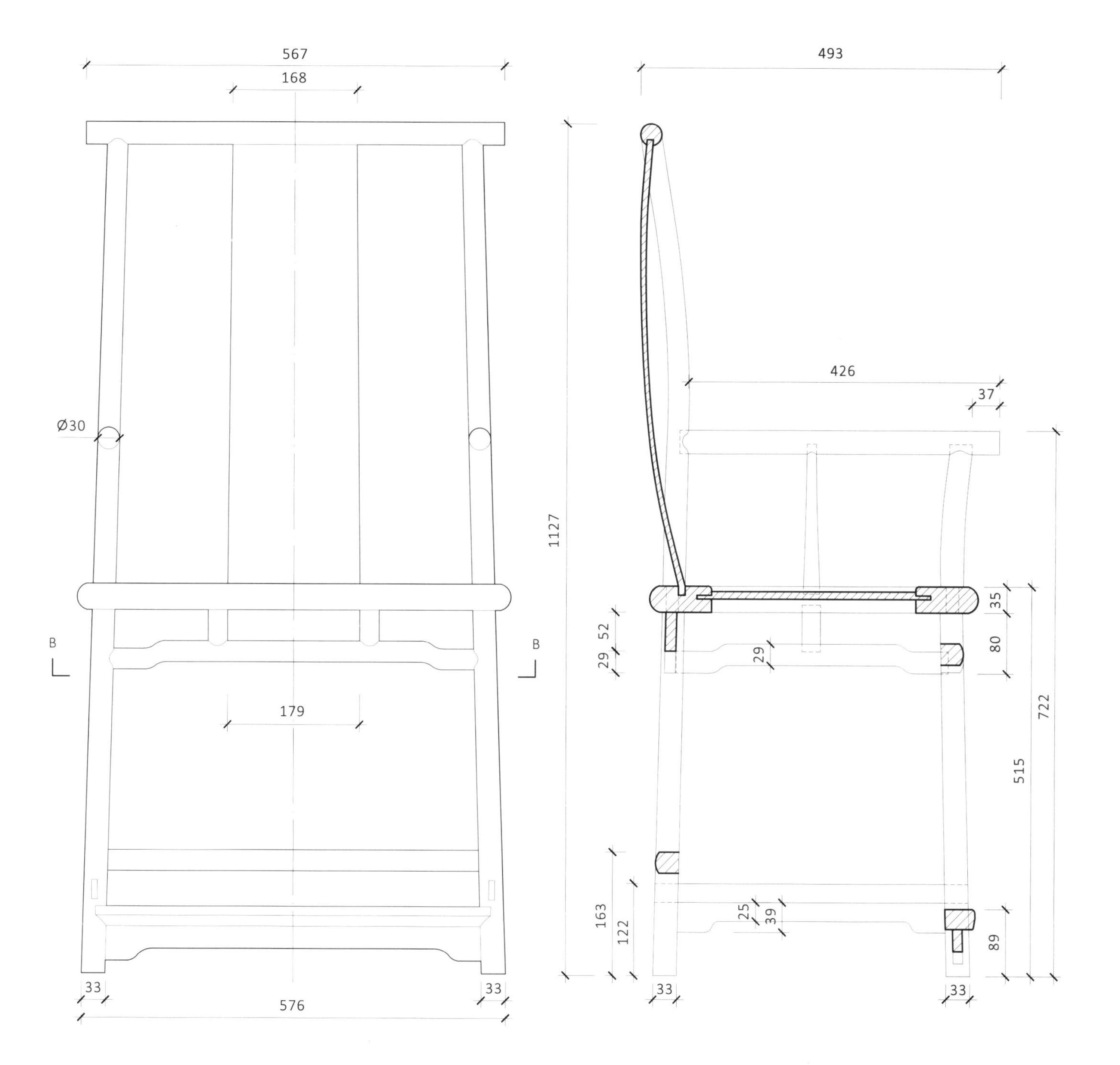

正视图 · FRONT ELEVATION ·

A剖面图 · SIDE AT A ·

杜丽娟 测绘
DU LIJUAN DELIN.

灯挂椅
Lamp-hanger chair

此椅原型为“明式灯挂椅”。灯挂椅是历史悠久的汉族椅类家具，因其造型好似南方挂在灶壁上用以承托油灯灯盏的竹制灯挂而得名。整体感觉是挺拔向上，简洁清秀，为明式家具的代表作。

规格
长 x 宽 x 高　490 x 428 x 1090（mm）

复原者
北京建筑大学 工业设计091班
指导教师：陈静勇 郎世奇

Its prototype is the Ming-style lamp-hanger chair. It is a very ancient furniture for Han people and it is named from the bamboo lamp-hanger which hang on the stove wall to support the oil lamp in the south of China. It is tall and straight, simple and elegant. The lamp-hanger chair is a magnum opus on the Ming-style furniture.

Dimension
length x width x height 490 x 428 x 1090 (mm)

Maker
Beijing University of Civil Engineering and Architecture
Industiral Design Class of 091
Advisory teachers:Chen Jingyong, Lang Shiqi

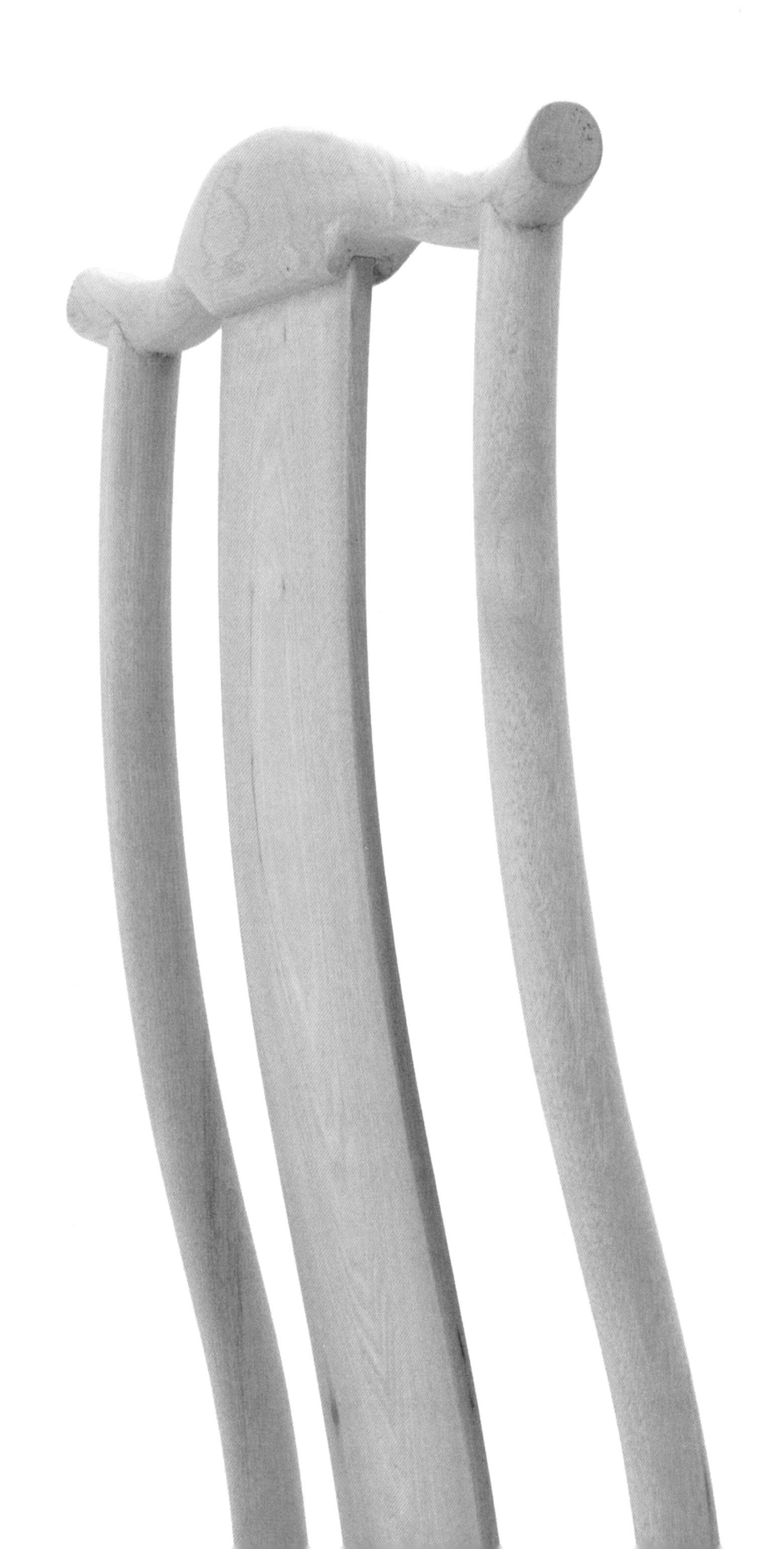

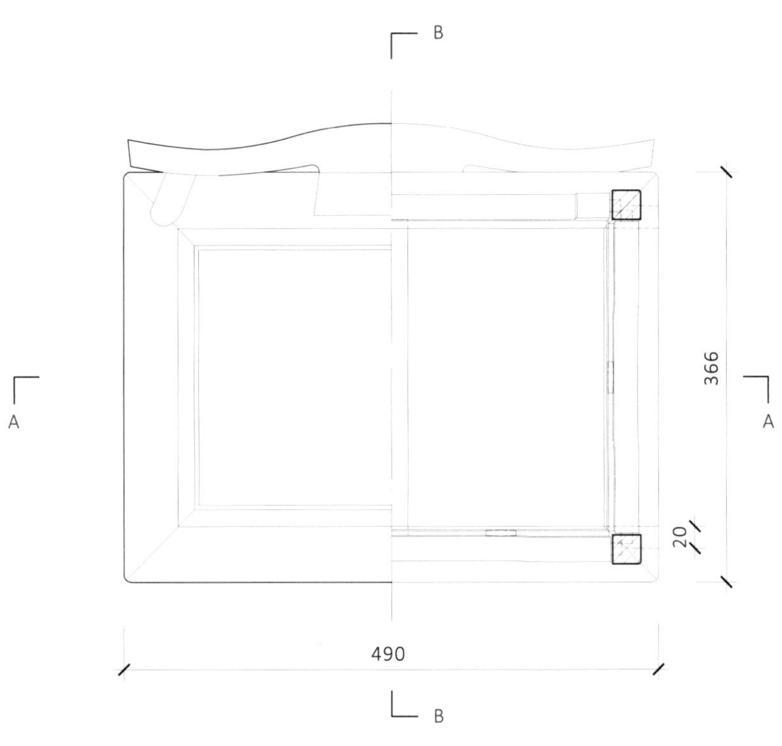

顶视图 · PLAN OF SEAT · C剖面图 · PLAN AT C ·

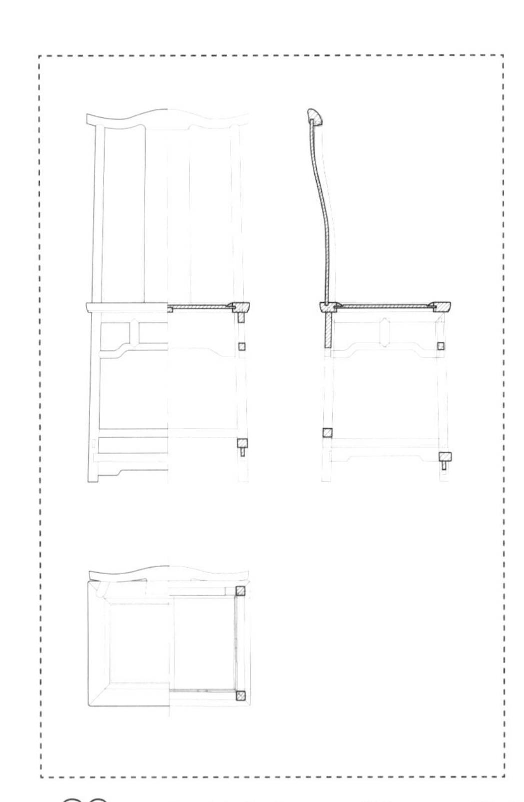

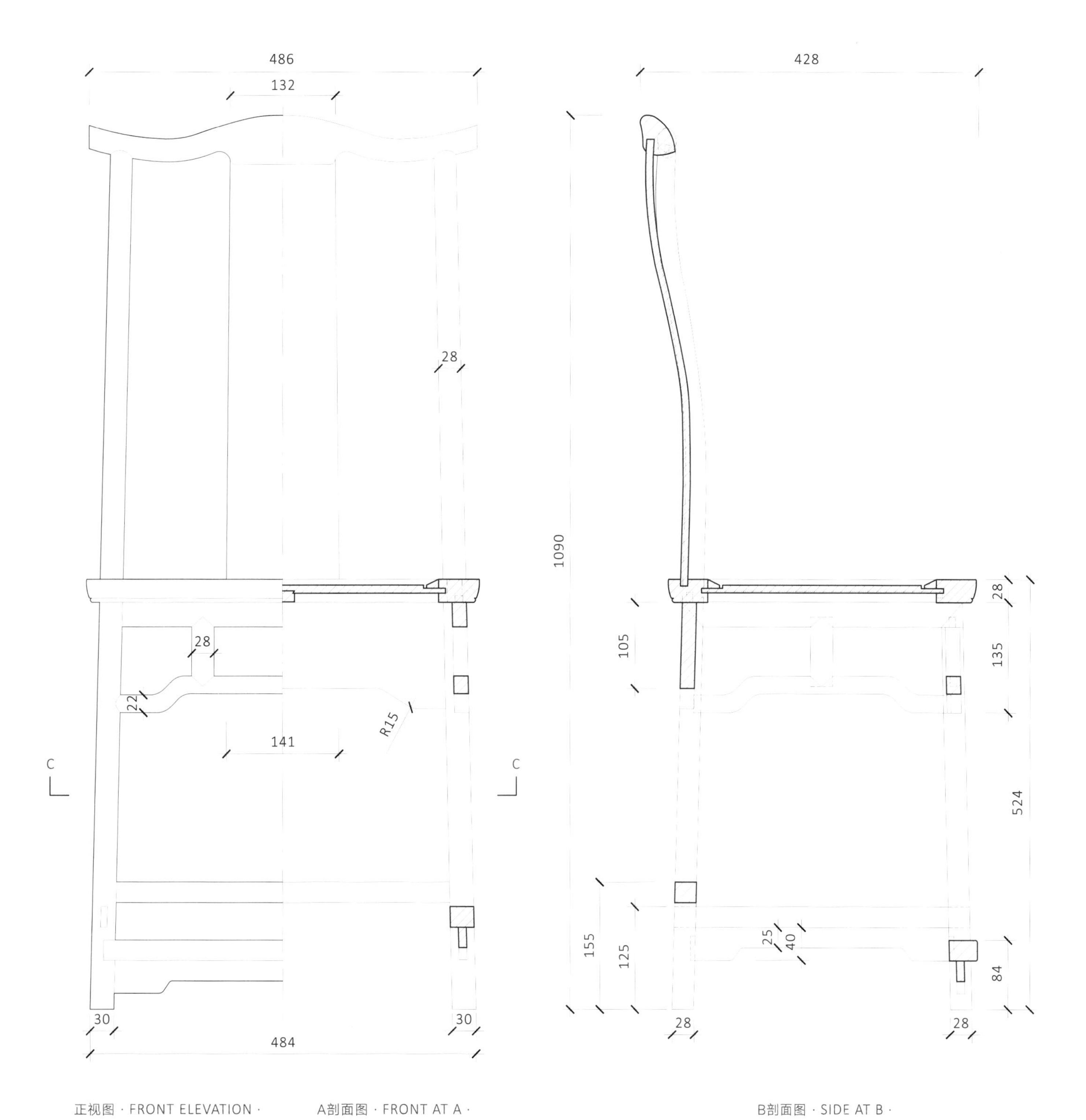

正视图 · FRONT ELEVATION · A剖面图 · FRONT AT A · B剖面图 · SIDE AT B ·

孔令奇 测绘
KONG LINGQI DELIN.

圈椅
Round-backed armchair

此椅原型为“明式素圈椅”。大体光素，椅盘以上为圆材，以下外圆内方。三面素牙条，除靠背板有浮雕、扶手下小角牙外，可以视作圈椅的基本形式。

规格
长 x 宽 x 高　704 x 605 x 1022（mm）

复原者
北京建筑大学 工业设计111班
指导教师：陈静勇 郎世奇 韩风 尚华

The prototype of the chair is "Round-backed armchair of Ming-style." It is neat and straightforward. Above the seat, the pole is round, other pole is round outside and square inside. It supported with spandrel on three sides. Apart from the reliefs on the back and the little spandrels under the arms, it shows the basic type of round-backed armchair.

Dimension
length x width x height 704 x 605 x 1022 (mm)

Maker
Beijing University of Civil Engineering and Architecture
Industiral Design Class of 111
Advisory teachers: Chen Jingyong, Lang Shiqi, Han Feng, Shang Hua

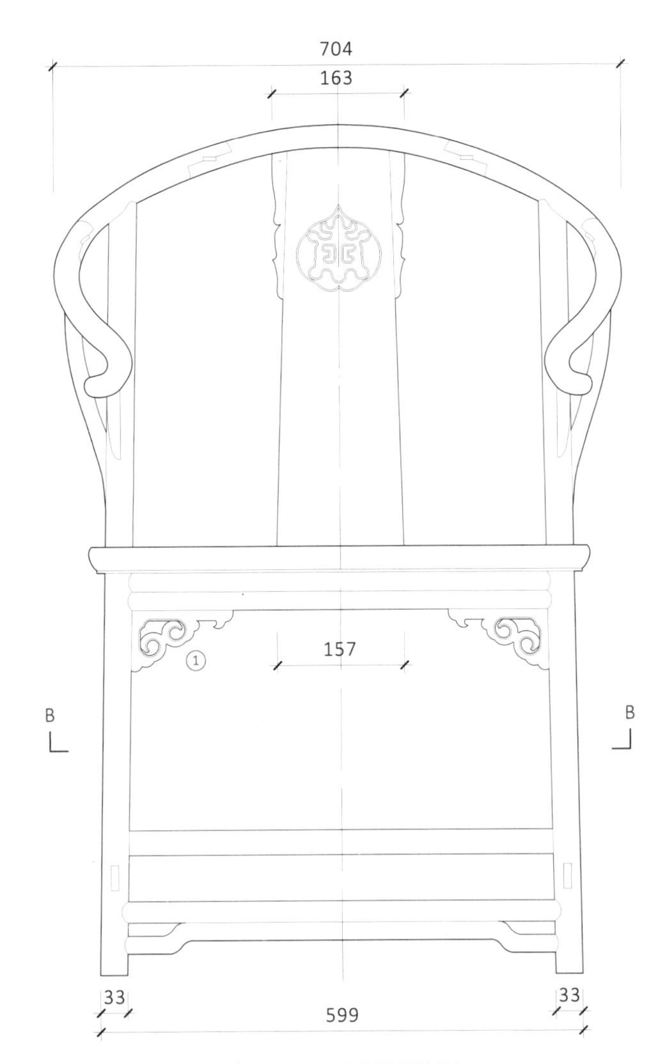

正视图 · FRONT ELEVATION ·

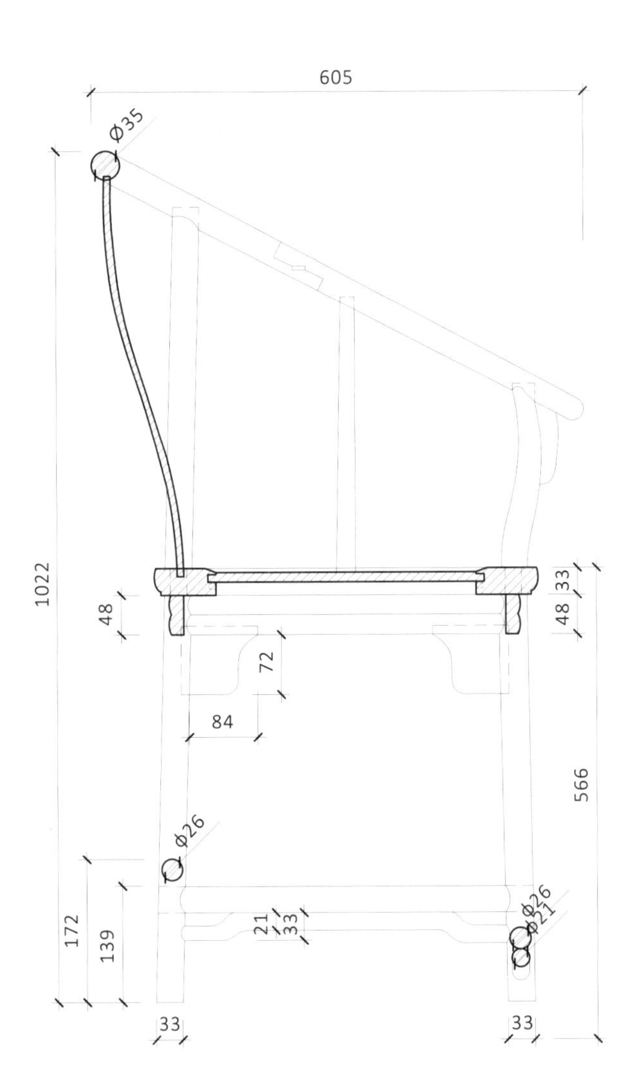

A剖面图 · SIDE AT A ·

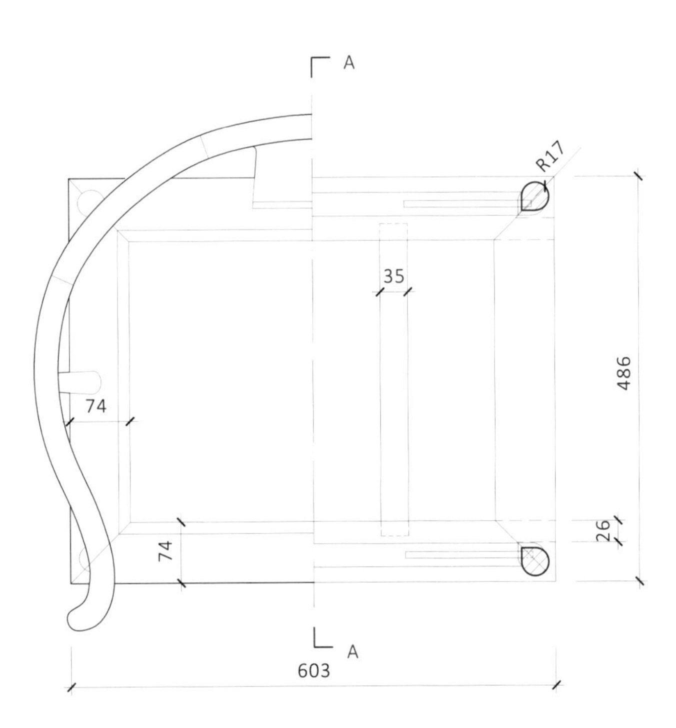

顶视图 · PLAN OF SEAT · B剖面图 · PLAN AT B ·

杜丽娟 测绘
DU LIJUAN DELIN.

六方形南官帽椅
Southern official's hat armchair

此椅原型为“明式六方形南官帽椅”。六方椅在明式家具中极为罕见，此椅尺寸甚至大于一般扶手椅，采用了较为复杂的线脚，是一个大胆的创新。此椅造型独特，虽是变体，但意趣清新，自然大方，无矫揉造作之感。

规格
长 x 宽 x 高　800 x 563 x 962（mm）

复原者
北京建筑大学 工业设计121班
指导教师：陈静勇 郎世奇

The prototype of the chair is southern officer's hat armchair with Ming style. It is very rare among the Ming-style furniture, and the size even larger than typical armchairs. It used a sophisticated skintle which is a innovative design. Although it is a variant of the original one, it is novel yet natural.

Dimension
length x width x height 800 x 563 x 962 (mm)

Maker
Beijing University of Civil Engineering and Architecture
Industiral Design Class of 121
Advisory teachers: Chen Jingyong, Lang Shiqi

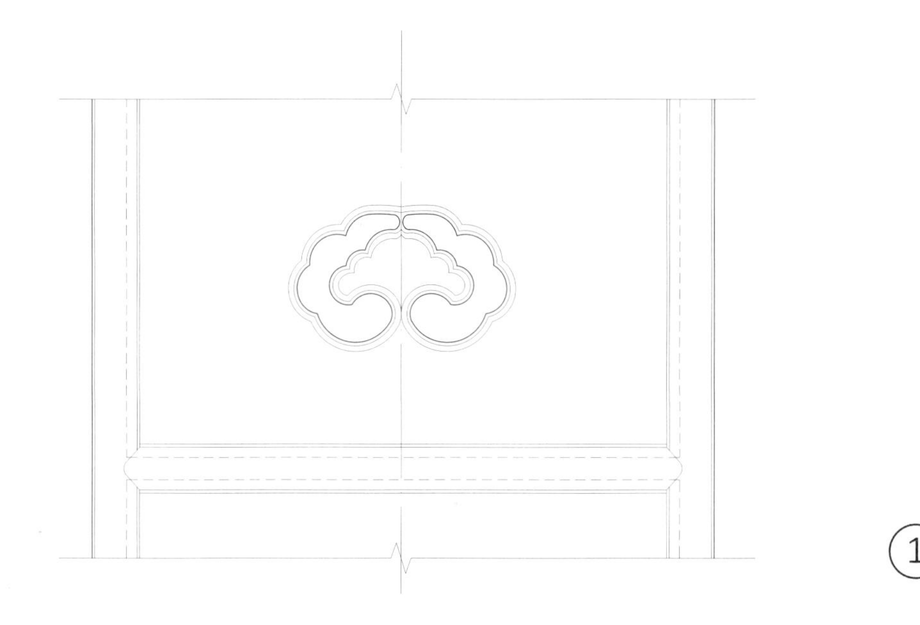
1

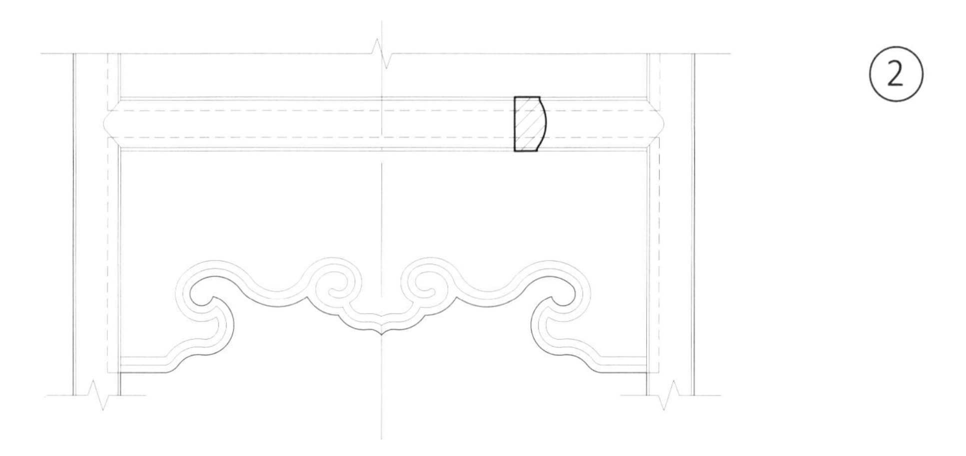
2

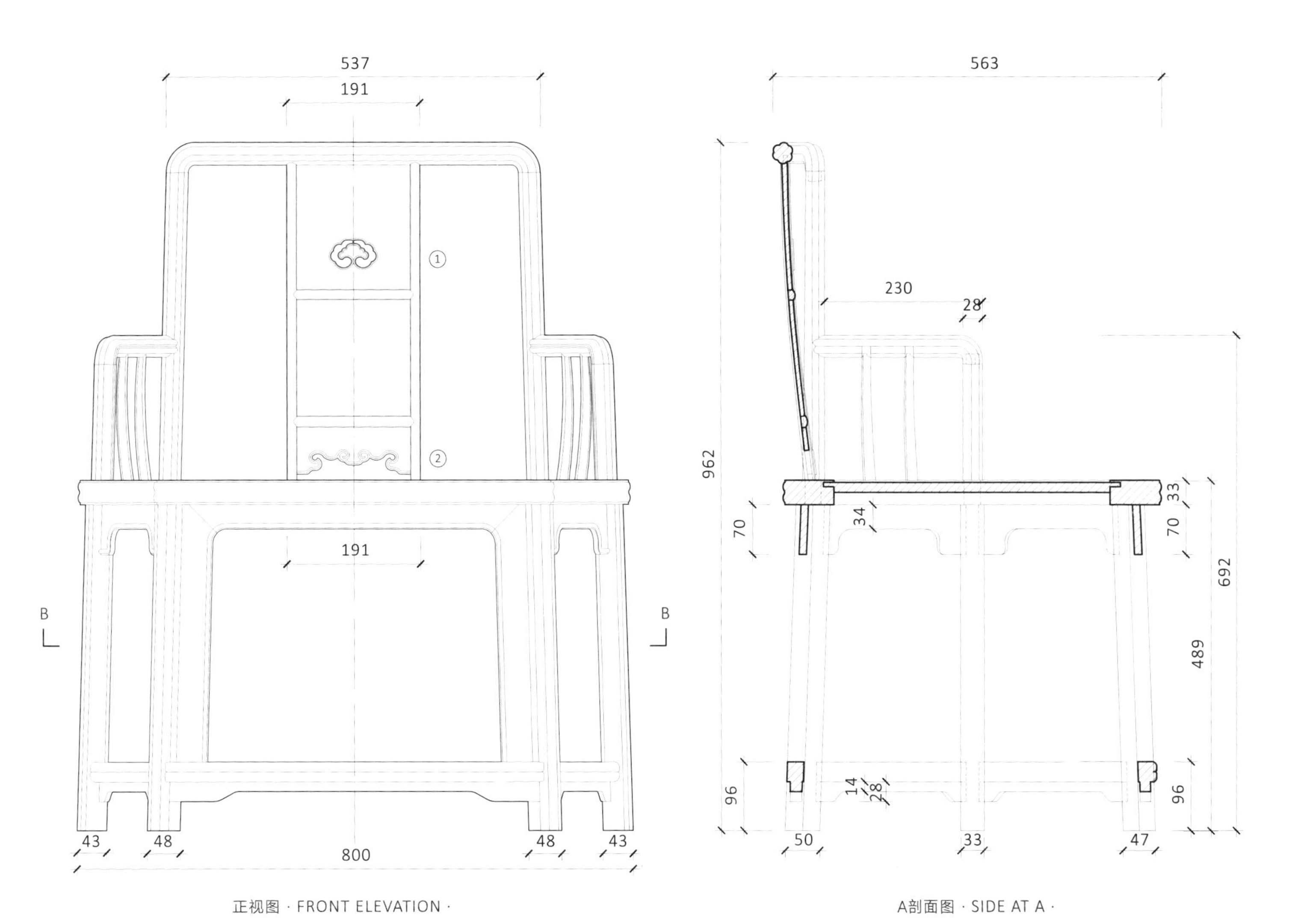

正视图 · FRONT ELEVATION ·

A剖面图 · SIDE AT A ·

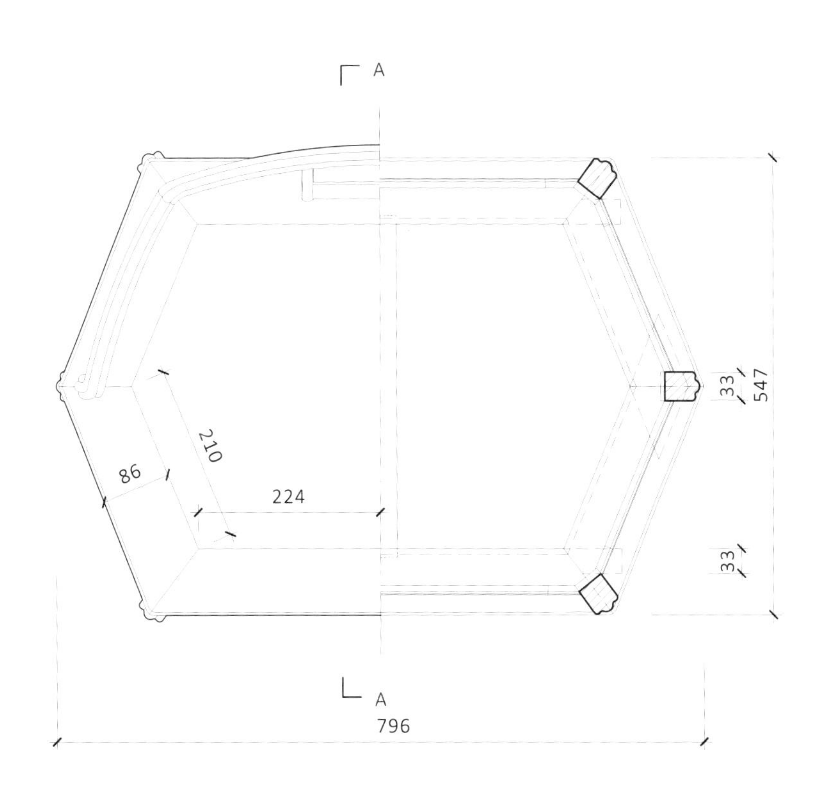

顶视图 · PLAN OF SEAT · B剖面图 · PLAN AT B ·

姬笑笑、张可凡、孔令奇 测绘
JI XIAOXIAO, ZHANG KEFAN and KONG LINGQI DELIN.

玫瑰椅
Rose chair

此椅原型为“明式券口靠背玫瑰椅”。玫瑰椅是明式扶手椅中常见的形式，其特点是靠背、扶手和椅面垂直相交，尺寸不大，用材较细，给人一种轻便灵巧的感觉。为明代各种椅子中较小的一种，用材单细，造型小巧美观。

规格
长 x 宽 x 高　612 x 460 x 828（mm）

复原者
北京建筑大学 工业设计121班
指导教师：陈静勇 郎世奇 尚华

The prototype of the chair is the rose chair with Ming style. The rose chair is a very typical Ming-style armchair. Its feature is that the back, the arm, and the seat are in vertical with each other. With a proper size and delicate material, the rose chair is one of the smallest chairs in Ming dynasty. It shows handy, smart and elegant fashion, giving the sense of lightweight and agility to people.

Dimension
length x width x height　612 x 460 x 828 (mm)

Maker
Beijing University of Civil Engineering and Architecture
Industiral Design Class of 121
Advisory teachers: Chen Jingyong, Lang Shiqi, Shang Hua

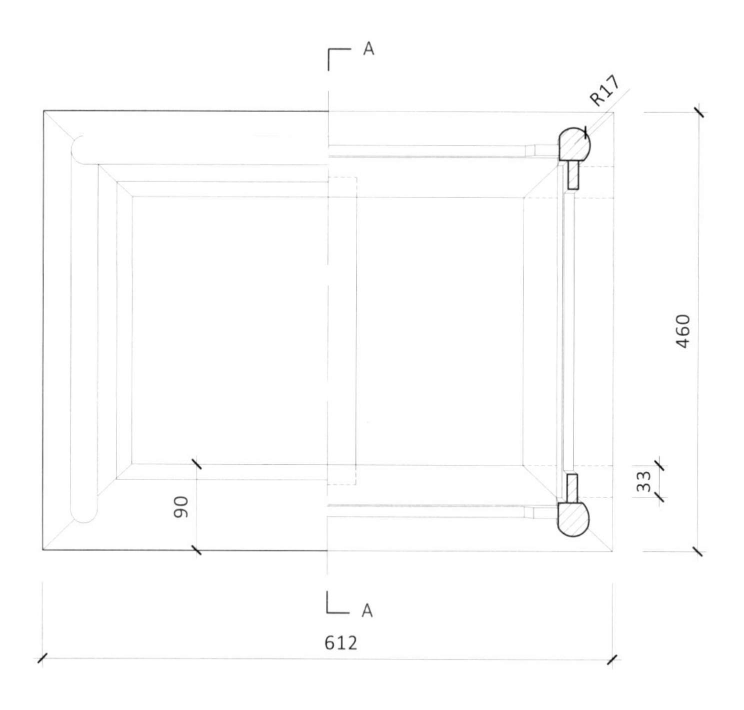

顶视图 · PLAN OF SEAT ·　　B剖视图 · PLAN AT B ·

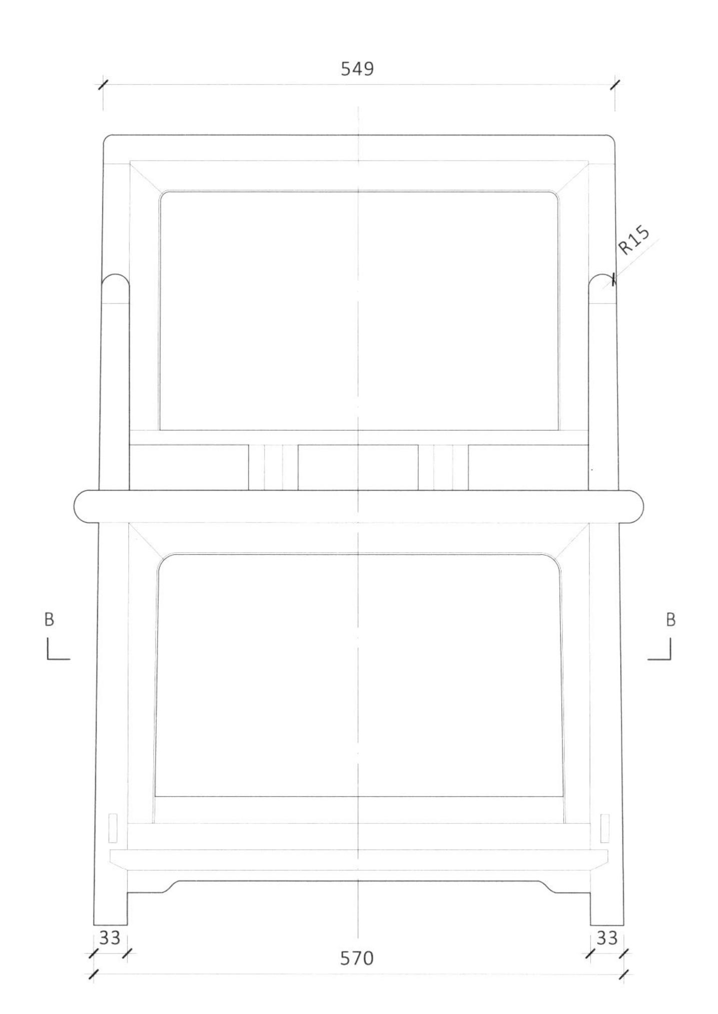

正视图 · FRONT ELEVATION ·

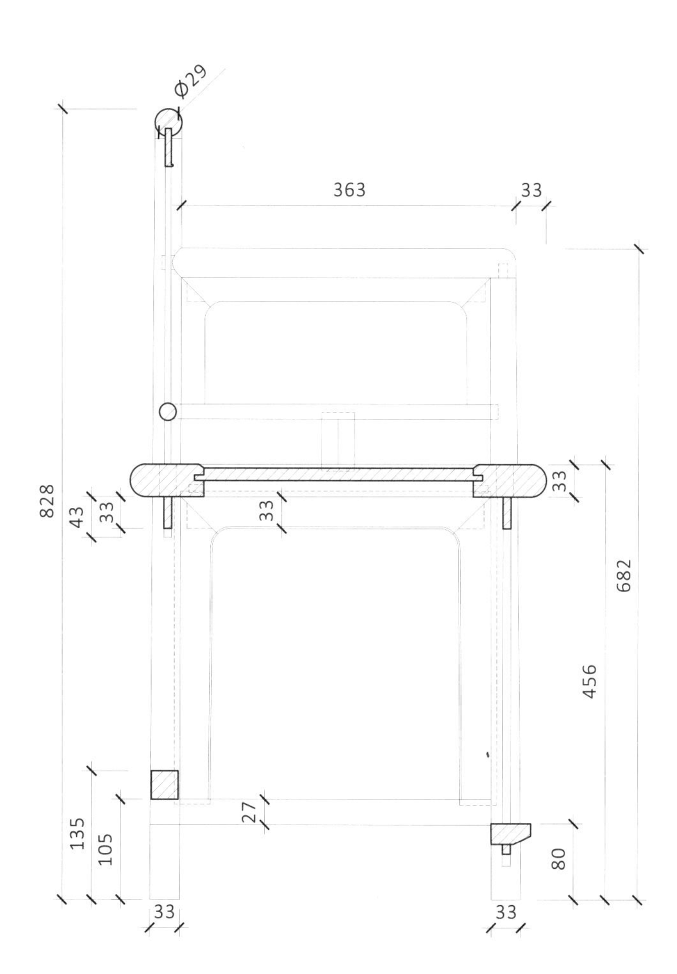

A剖面图 · SIDE AT A ·

潘阳 测绘
PAN YANG DELIN.

一腿三牙罗锅枨方桌
Three spandrels humpbacked stretcher Square table

此椅原型为“明式黄花梨一腿三牙罗锅枨方桌”。此种方桌是明式方桌中的一种常见形式。腿足及罗锅枨上犀利有力的锐棱峭拔精神，使方桌显得骨相清奇，劲挺不凡。

规格
长 x 宽 x 高　980 x 980 x 880（mm）

复原者
北京建筑大学 工业设计121班
指导教师：陈静勇 郎世奇

Its prototype is the Ming-style Huanghuali-wood three spandrels hump-backed stretcher square table, which is very common in Ming-style square table. The sharp and brisk lines on the leg and the hump-baked stretcher give the table a refreshing and strong spirit.

Dimension
length x width x height 980 x 980 x 880(mm)

Maker
Beijing University of Civil Engineering and Architecture
Industiral Design Class of 121
Advisory teachers: Chen Jingyong, Lang Shiqi

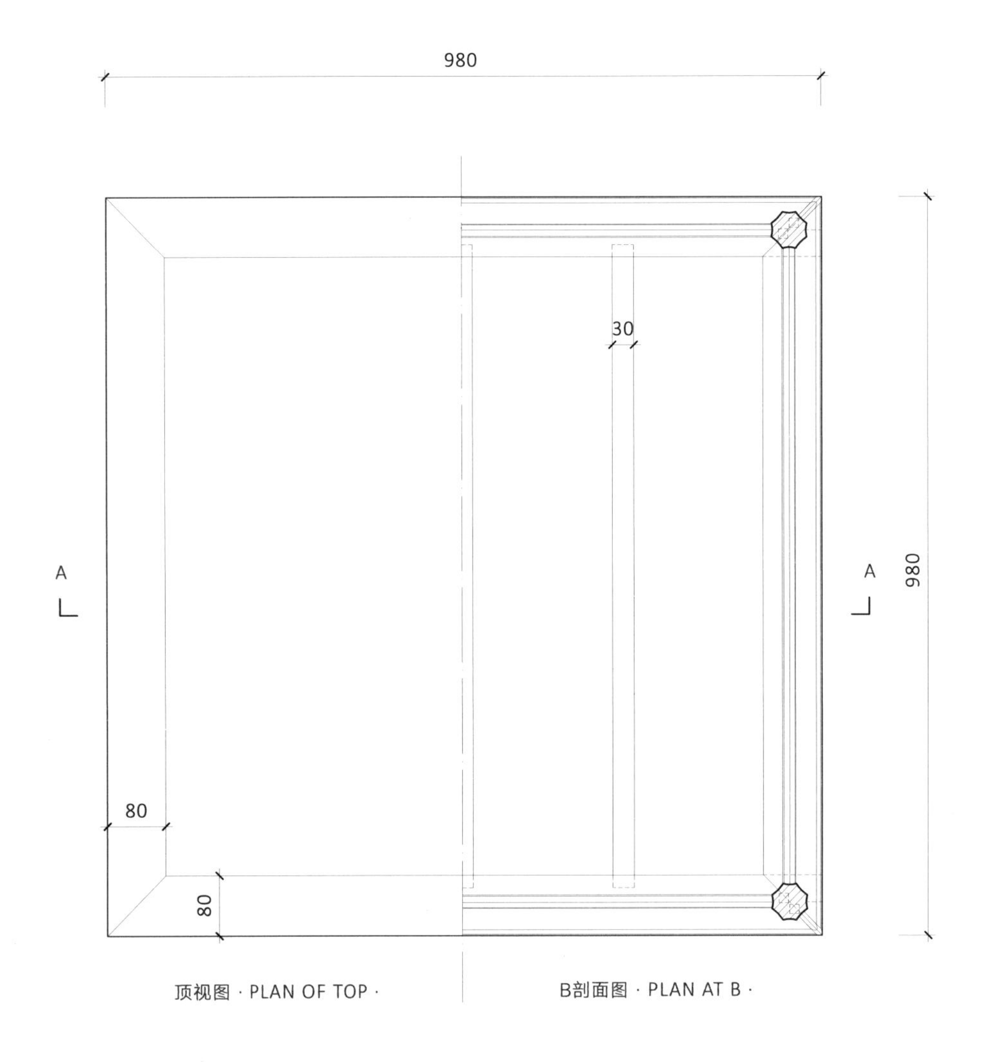
980
30
A
A
980
80
80
顶视图 · PLAN OF TOP ·
B剖面图 · PLAN AT B ·

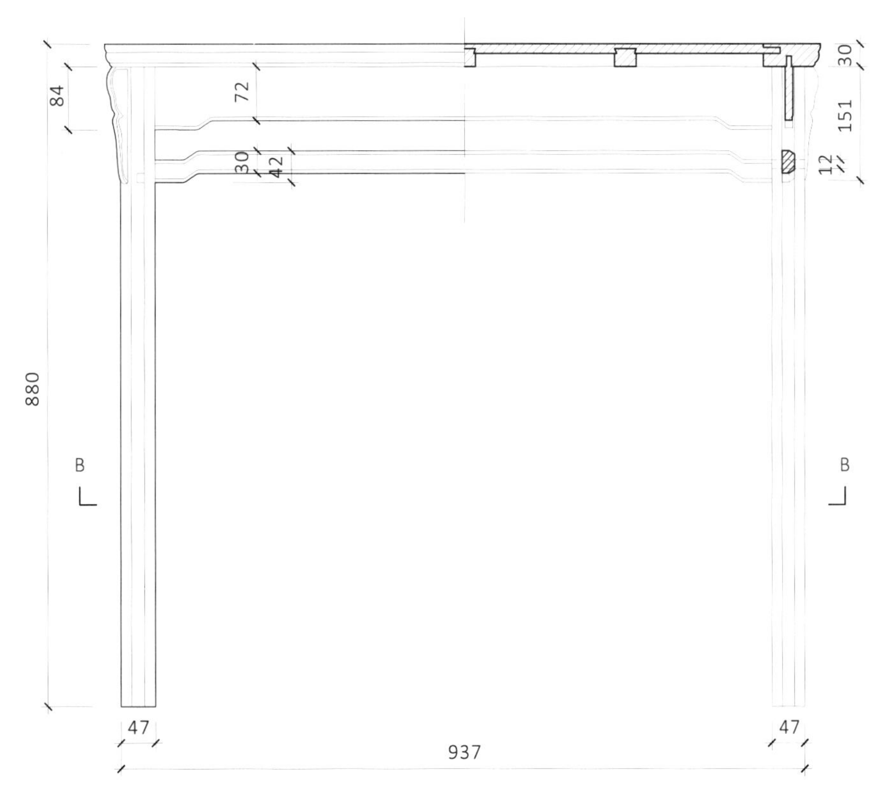

正视图 · FRONT ELEVATION ·　　A剖面图 · FRONT AT A ·

范雨佳 测绘
FAN YUJIA DELIN.

夹头榫带屉板平头案
Elongated bridle joint flat-top recessed-leg table with drawers

此案原型为“明式夹头榫带屉板平头案”。此带屉板平头案为典型平头案的变体。为了增加条案的使用空间与面积，案面下有一屉板，镶入腿足间枨子的槽口，枨子作肩接入腿足。

规格

长 x 宽 x 高　707 x 371 x 790（mm）

复原者

北京建筑大学-工业设计101班

指导教师：陈静勇 郎世奇

The prototype of the table is " Ming-style Elongated bridle joint flat-top recessed-leg table with drawers." This work is a variant of the classic flat-top table. To increase the space for use, there is the drawer under the table top. It is edged into the notched stretcher. The stretcher, as a shoulder, is then attached to the leg.

Dimension

length x width x height 707 x 355 x 790 (mm)

Maker

Beijing University of Civil Engineering and Architecture

Industiral Design Class of 101

Advisory teachers: Chen Jingyong, Lang Shiqi

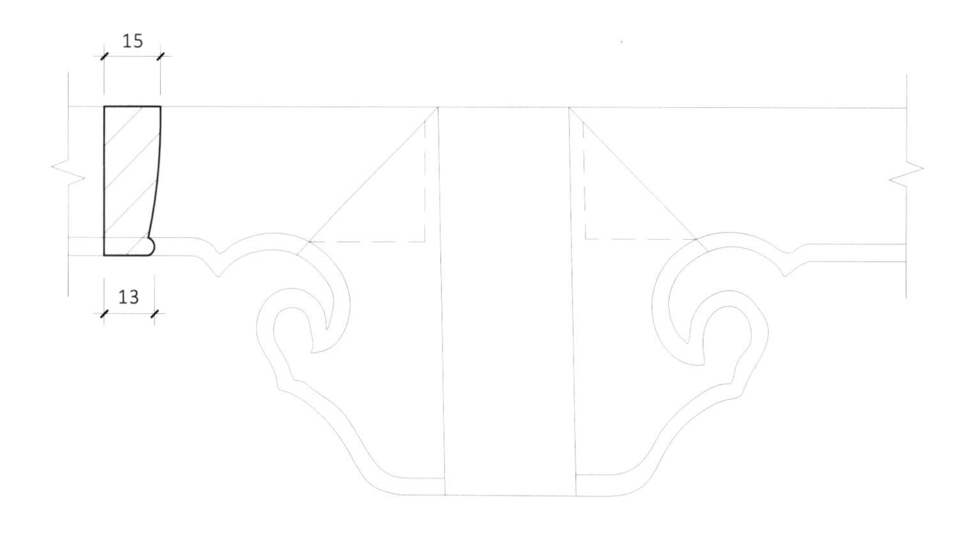
15
13
1

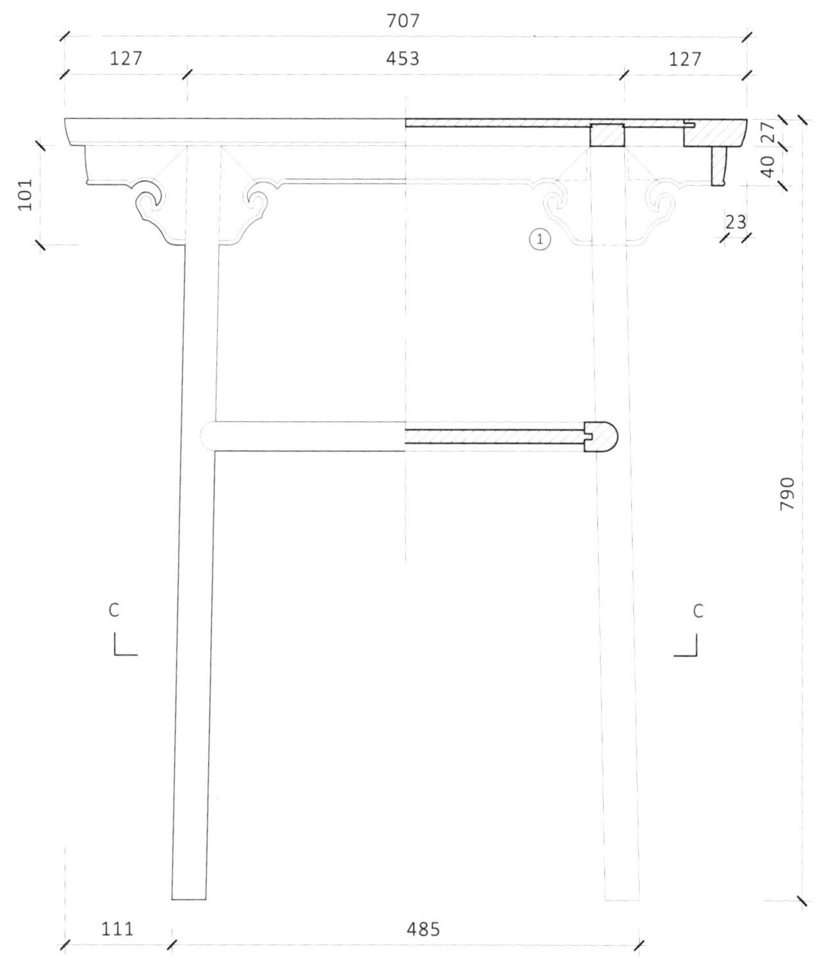

正视图 · FRONT ELEVATION · A剖面图 · FRONT AT A ·

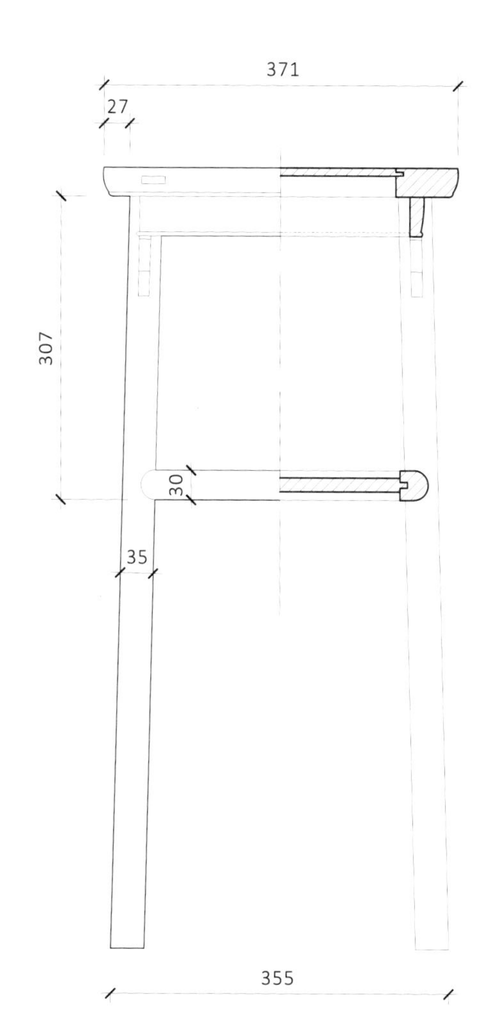

侧视图 · SIDE ELEVATION · B剖面图 · SIDE AT B ·

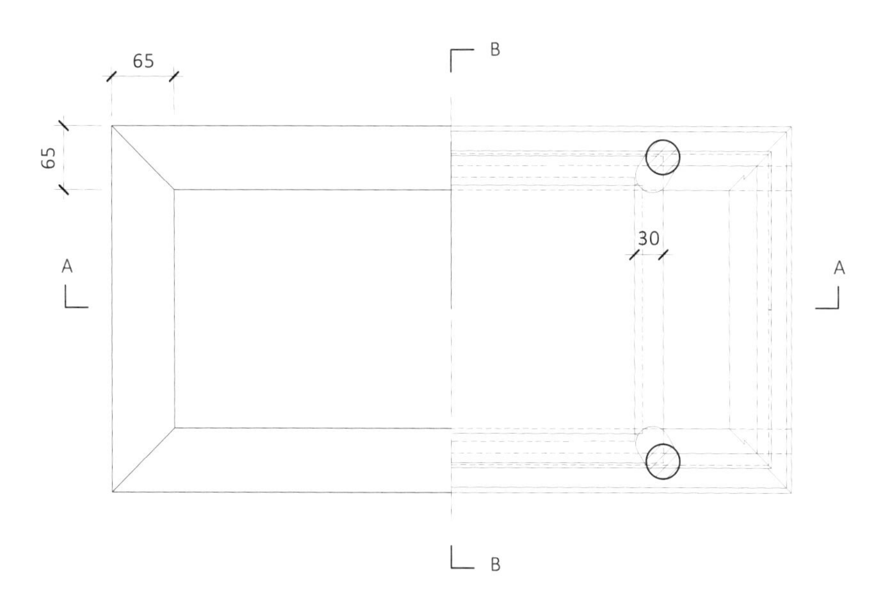

顶视图 · PLAN OF TOP · C剖面图 · PLAN AT C ·

赵帅 测绘
ZHAO SHUAI DELIN.

翘头案
Head desk

此案原型为“明式夹头榫带托子翘头案”。翘头案案面两端上翘，俗称飞角。翘头增加了案子的雅致轻盈，不失尚礼文秀。在中国古代常置于厅堂，造型灵巧庄重，是屋内器物陈设装饰的重要承具。

规格

长 x 宽 x 高　1400 x 463 x 836（mm）

复原者

北京建筑大学 环境设计131班

指导教师：陈静勇　郎世奇　韩风　尚华

The prototype of the table is " Ming-style Elongated bridle joint everted flanges Recessed-leg table with Tray." Both ends of the table are up. It is commonly known as flying horn. Everted flanges increase the sense of elegance and lightness of table, and not lost gentle and pretty. It always set in the hall in ancient China. It is a significant table which shape is dexterity and solemnity with objects placed in the house.

Dimension

length x width x height 1400 x 463 x 836 (mm)

Maker

Beijing University of Civil Engineering and Architecture
Environment Design Class of 131
Advisory teachers: Chen Jingyong, Lang Shiqi, Han Feng, Shang Hua

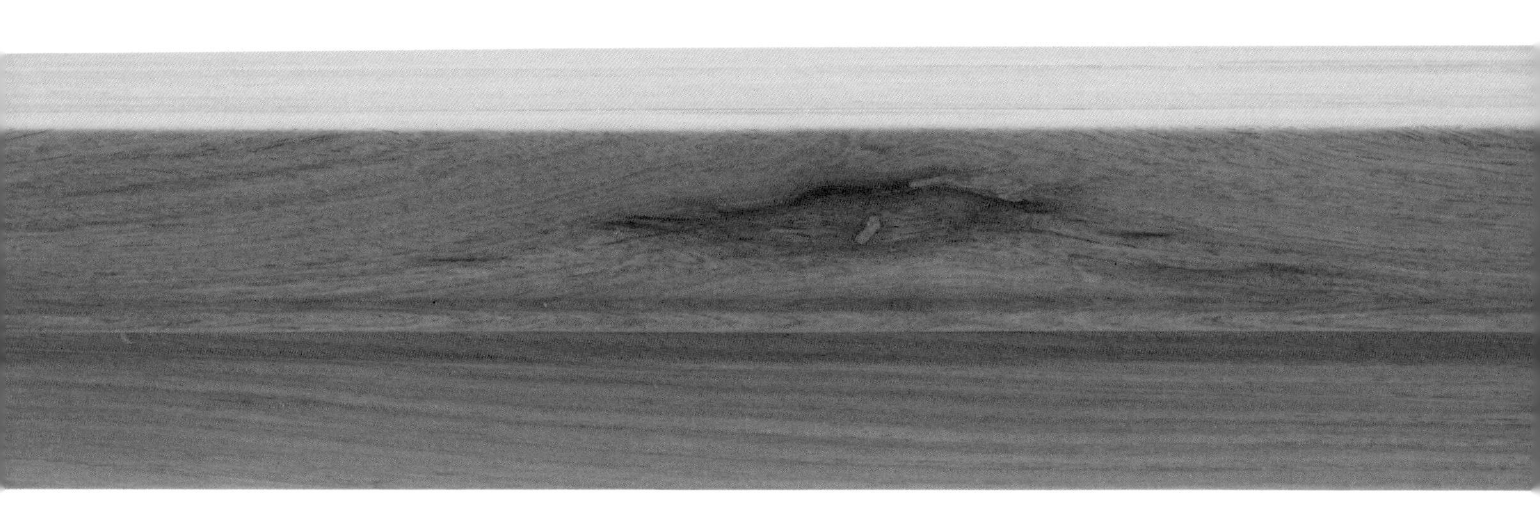

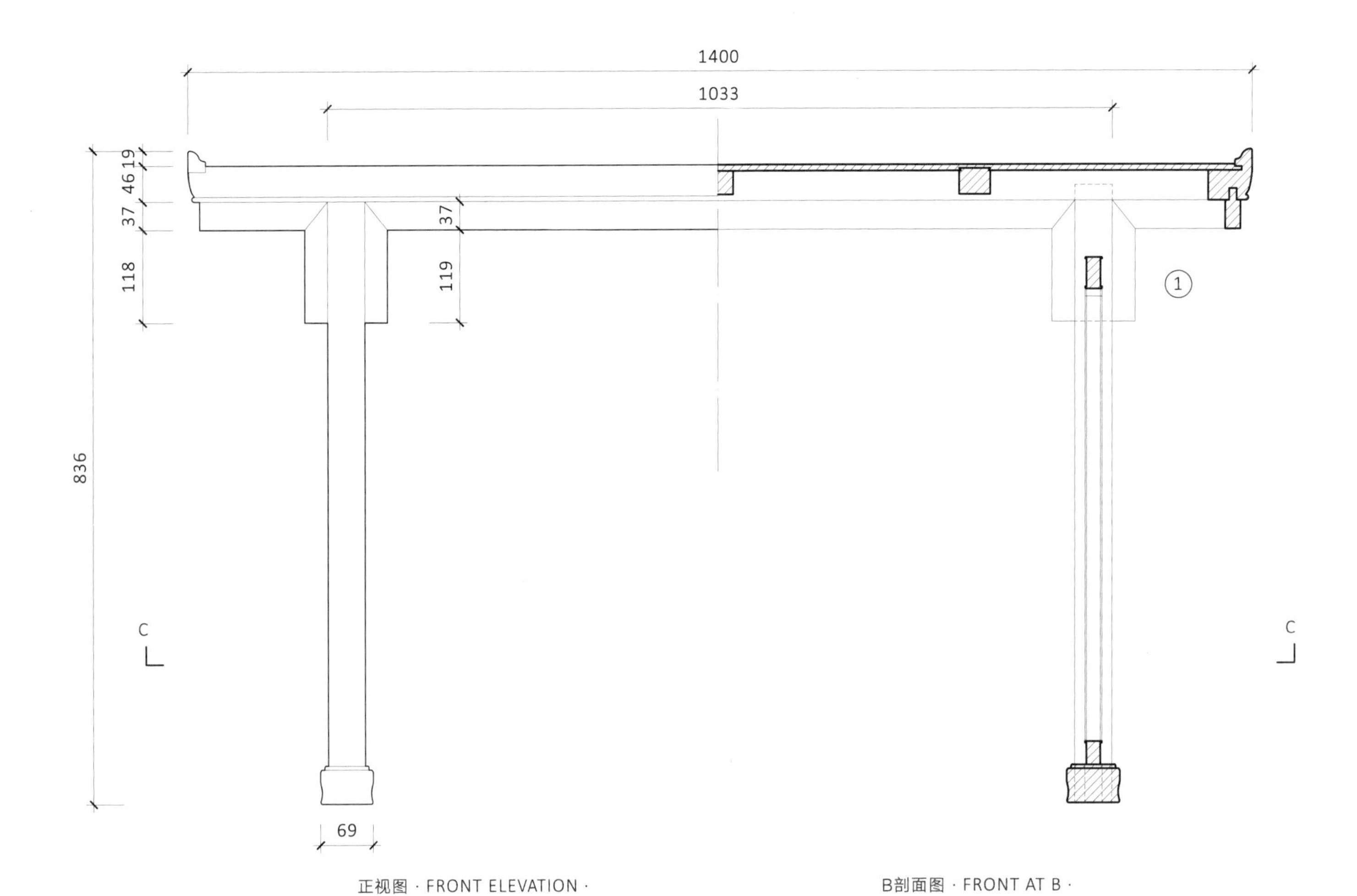

正视图 · FRONT ELEVATION ·

B剖面图 · FRONT AT B ·

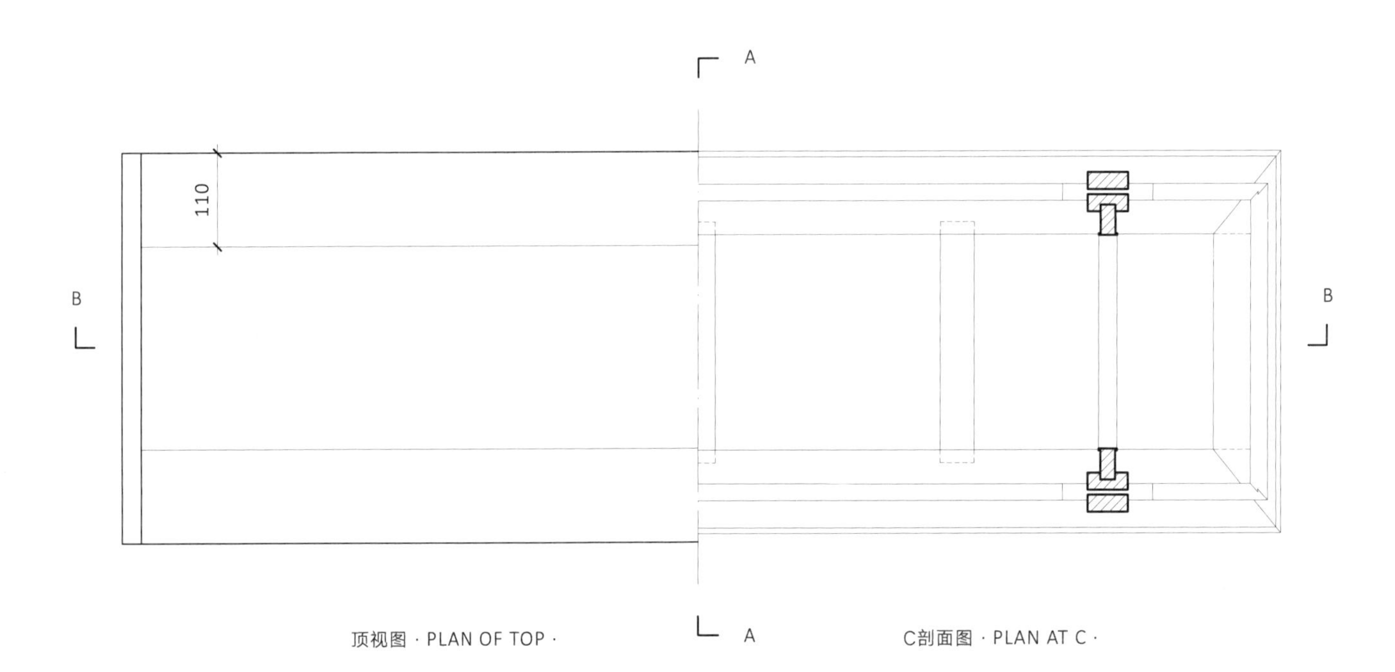

顶视图 · PLAN OF TOP ·

C剖面图 · PLAN AT C ·

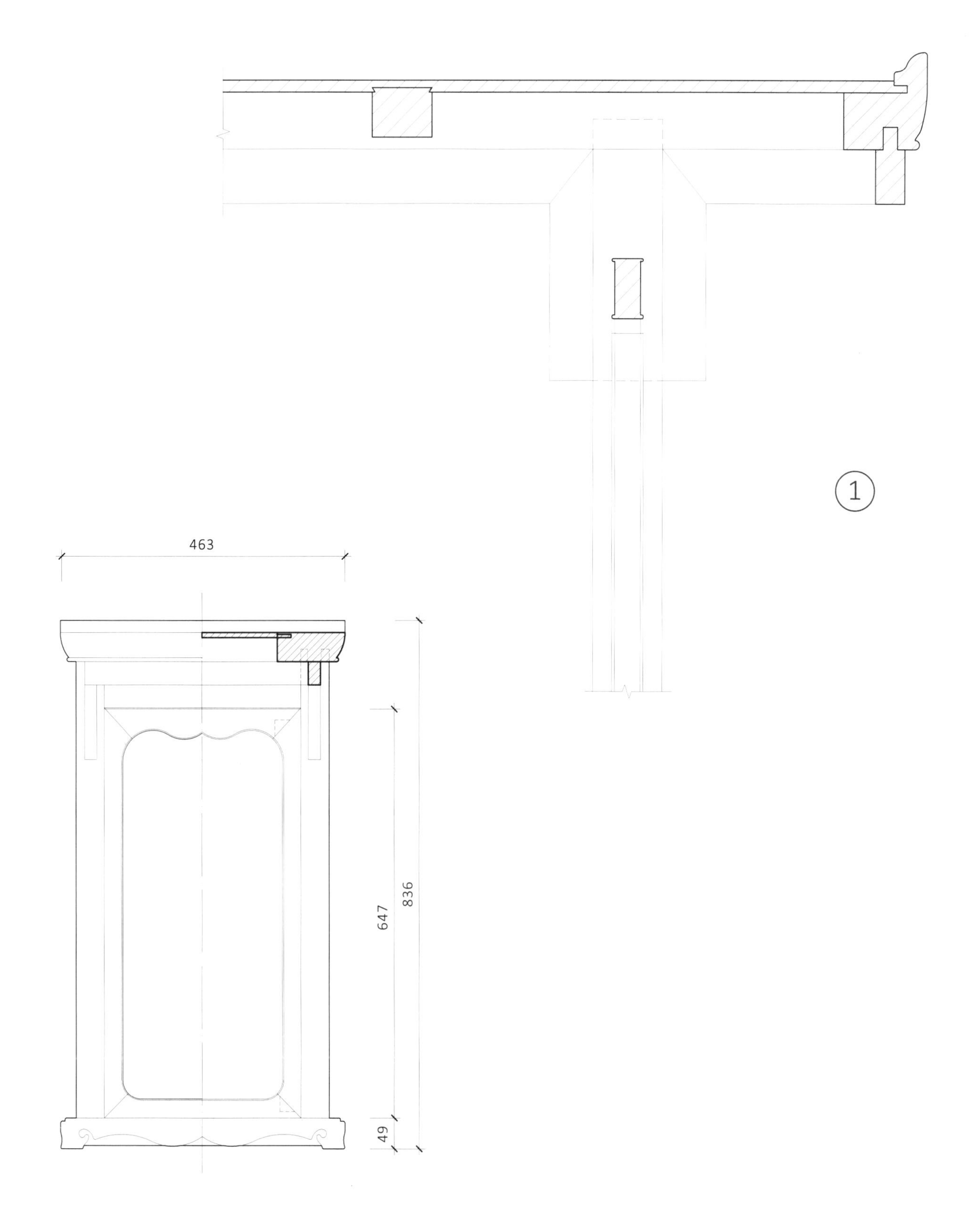

侧视图 · SIDE ELEVATION ·　　　　A剖面图 · SIDE AT A ·

林则全、张晓丽 测绘
LIN ZEQUAN, ZHANG XIAOLI DELIN.

总统桌
President table

此总统桌的设计诠释了“中国设计，葡国工艺”，为纪念总统会见“驻场计划”北京设计师代表团而设计。由桌面、桌腿、托泥3个部位组成，部件间通过螺栓连接，形成梯形框架形态；采用当地胡桃木制作，极简的线条梯形结构和材质的美感。

规格 长 x 宽 x 高 2500x800x785(mm)
设计者 杨琳
制作商 Guarnição, Lda

This table embodies Chinese design and Portugal crafts. It is designed to mark Portugal President's meeting with Beijing designers in the "Artist-In-Residence" program. This design is composed of the table top, legs and leg support. And the conjunction is screw bolt to shape the trapezoid structure. It made of local walnut wood which reflect the minimalist linear-trapezoid structure and the aesthetic material.

Dimension length x width x height 2500 x 800 x 785(mm)
Deigner Yang Lin
Maker Guarnição, Lda

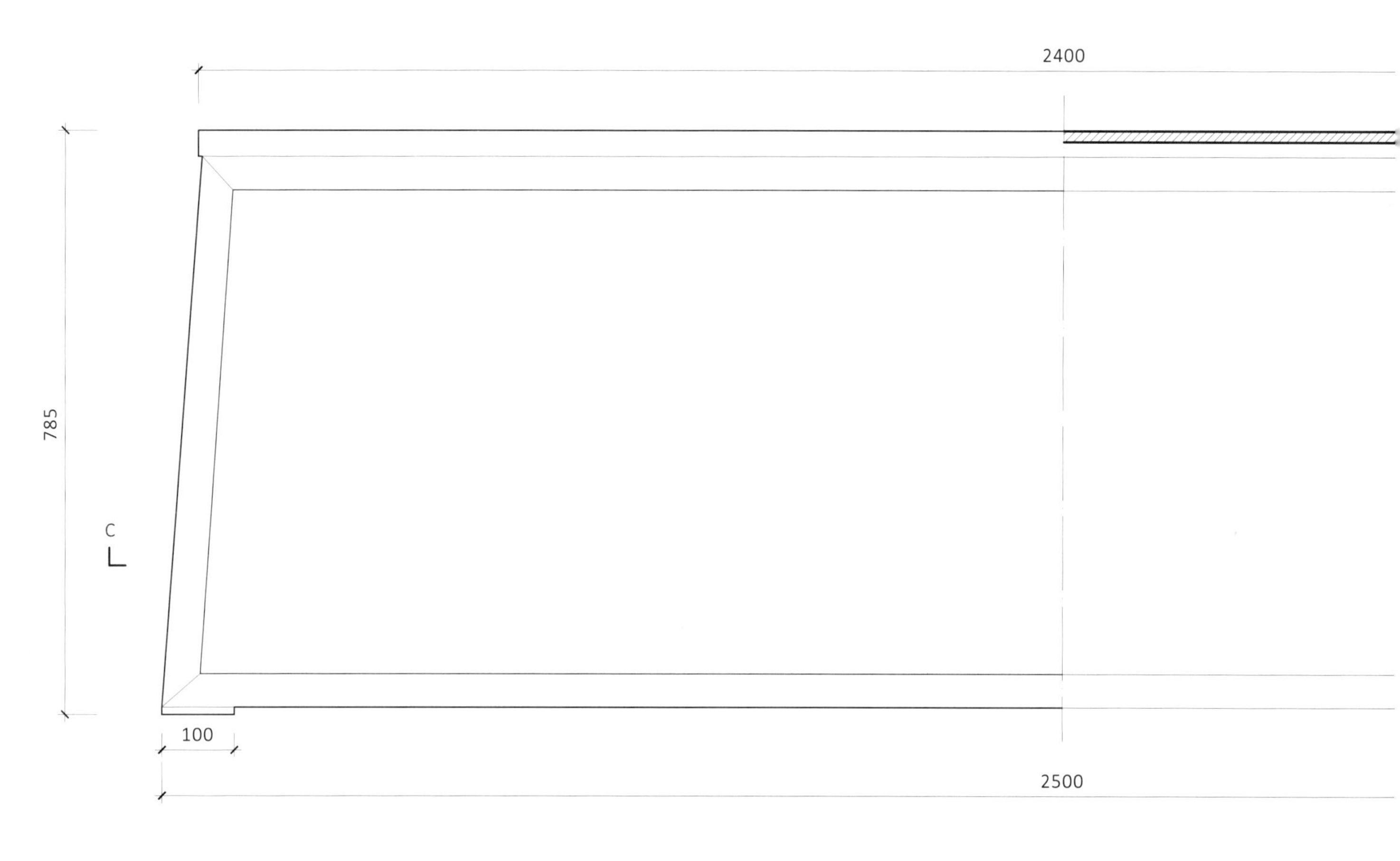

正视图 · FRONT ELEVATION ·

A剖面图 · FRONT AT A ·

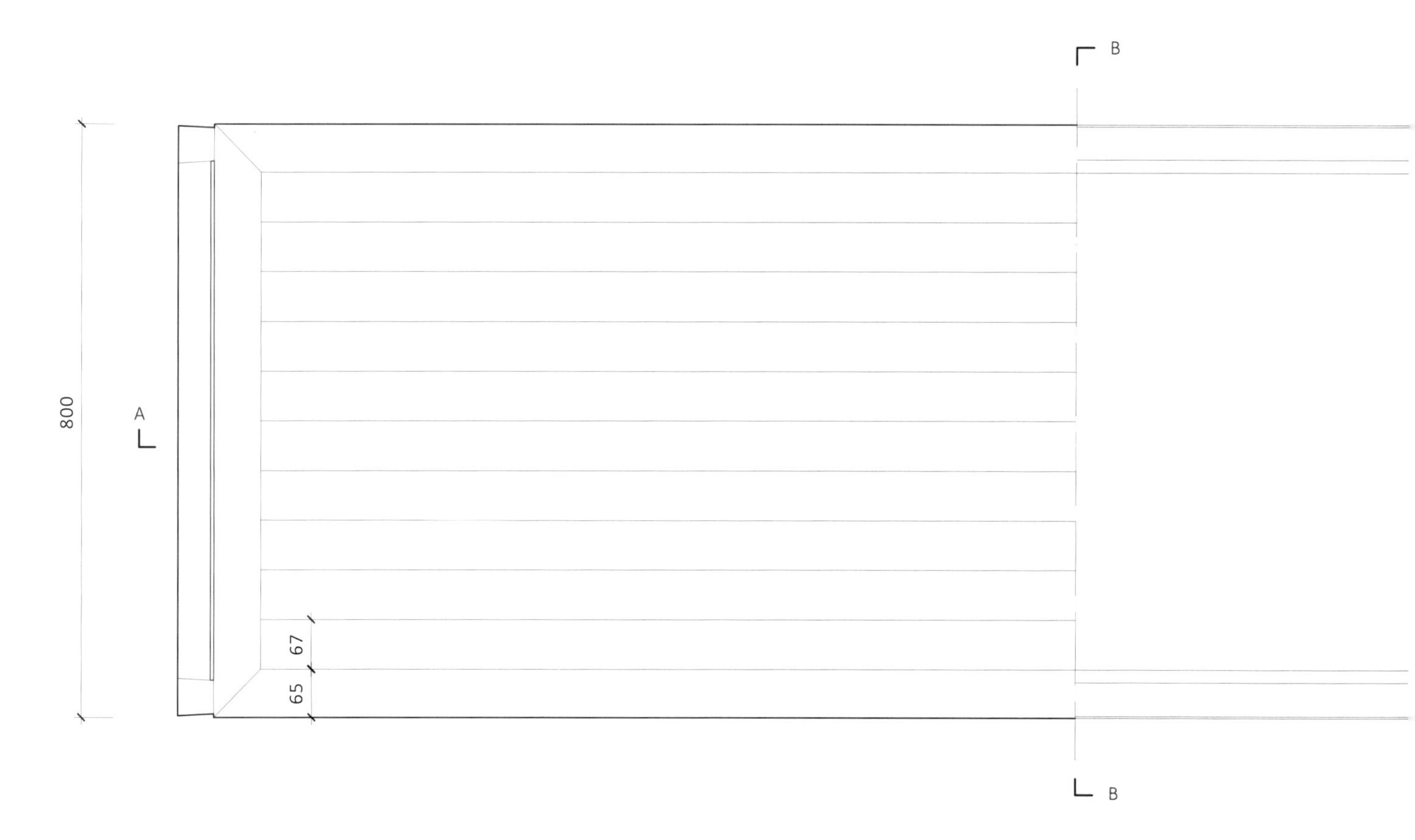

顶视图 · PLAN OF TOP ·

C剖面图 · PLAN AT C ·

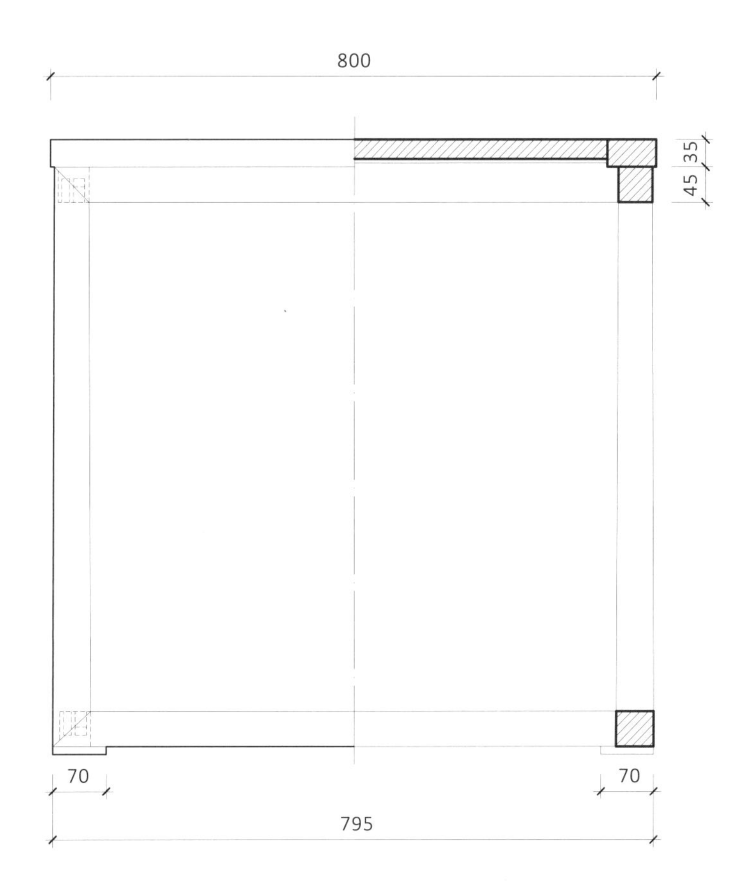

侧视图 · SIDE ELEVATION ·　　　　B剖面图 · SIDE AT B ·

刘泽 测绘
LIU ZE DELIN.

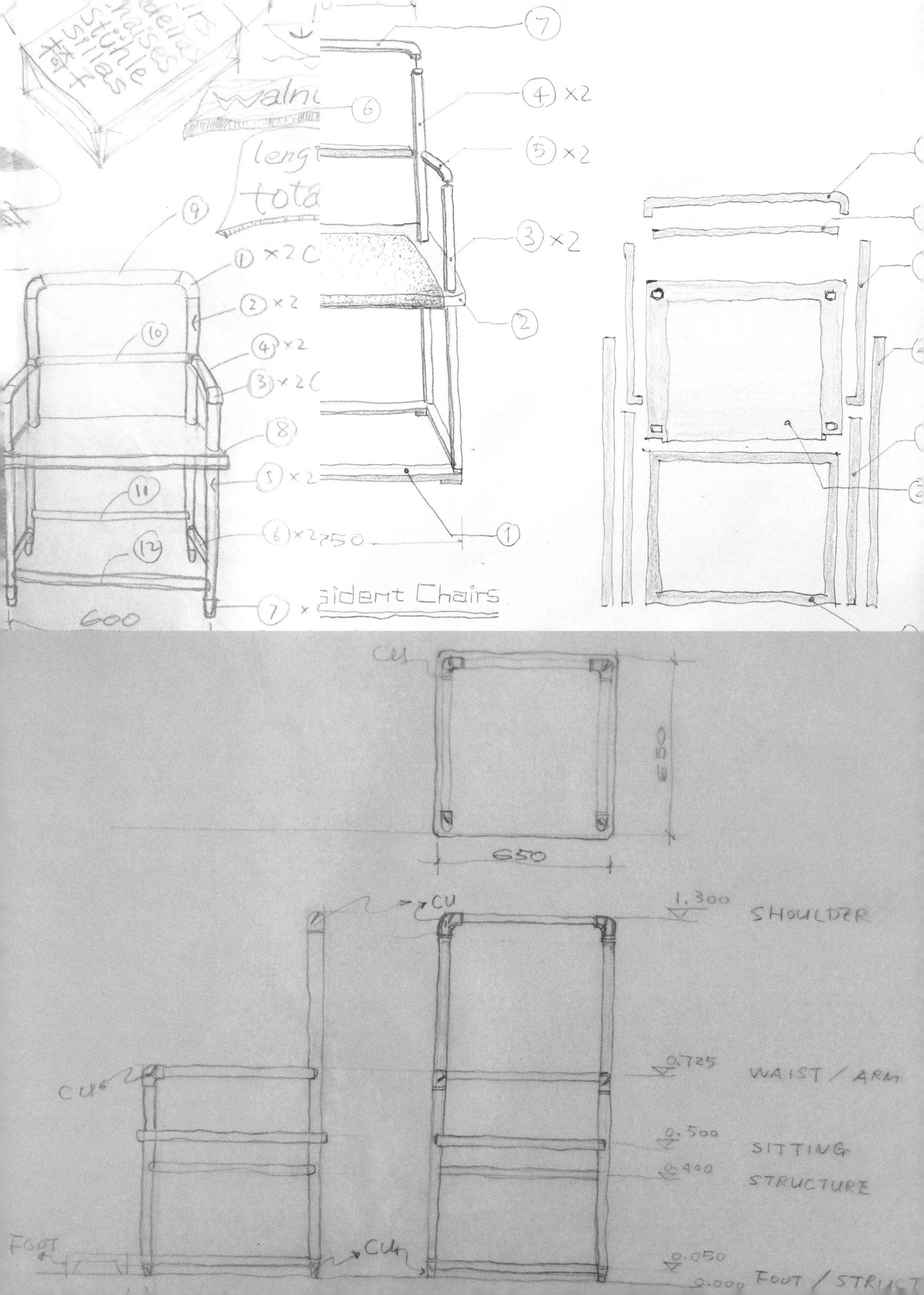

walnu
lengt
tota
600
ident Chairs
650
650
CU
1.300
SHOULDER
0.725
WAIST / ARM
0.500
SITTING
0.400
STRUCTURE
FOOT
0.050
0.000
FOOT / STRUCT

总统椅
President chair

总统椅的设计诠释了“中国设计，葡国工艺”，为纪念总统会见“驻场计划”北京设计师代表团而设计。极简的线条体现结构和材质的美感，采用胡桃木制做，椅面铺有软木；作品由椅腿、坐凳和椅背扶手3个独立部分组成，通过螺栓连接，方便独立自主拼装、运输。

规格 长 x 宽 x 高 750x750x1070(mm)
设计者 邹乐
制作商 Guarnição, Lda

This chair embodies Chinese design and Portugal crafts. It is designed to mark Portugal President's meeting with Beijing designers in the "Artist-In-Residence" program. The primary structure of the chair made of walnut wood. Its seat is covered with cork wood. This design is composed of the legs, the seat and the back and arm. They connected with bolts, convenient for assembling and transportation.

Dimension length x width x height 750 x 750 x 1070(mm)
Deigner Zou Le
Maker Guarnição, Lda

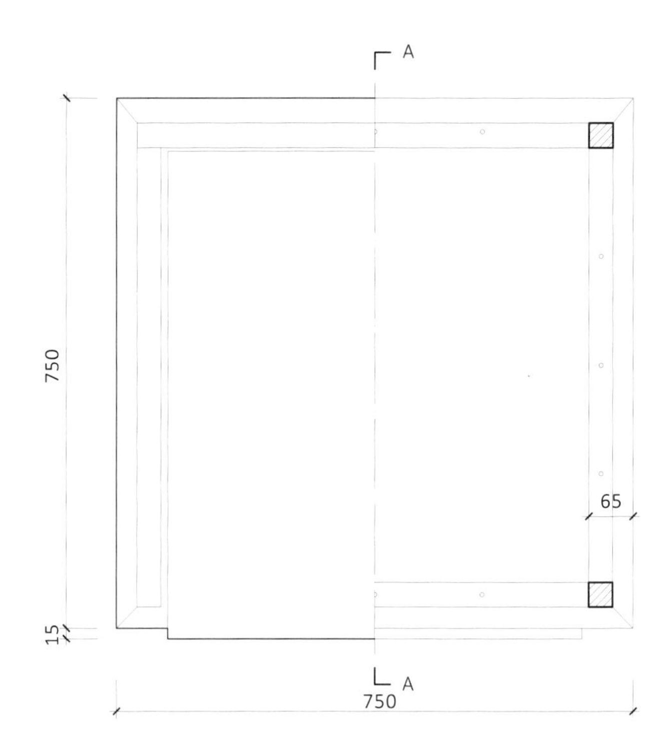

顶视图 · PLAN OF TOP · B剖面图 · PLAN AT B ·

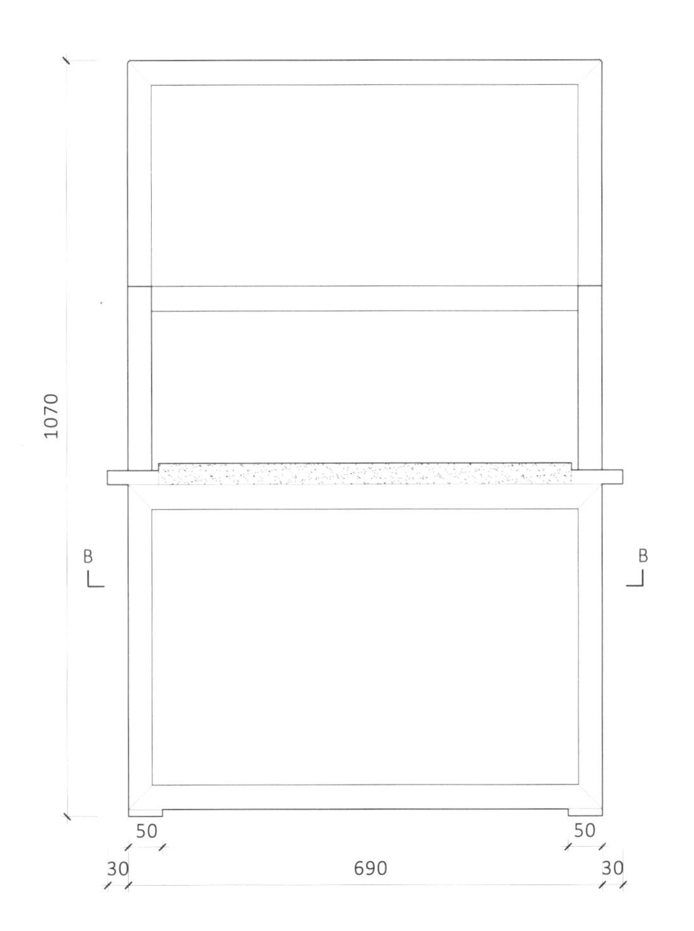

正视图 · FRONT ELEVATION ·

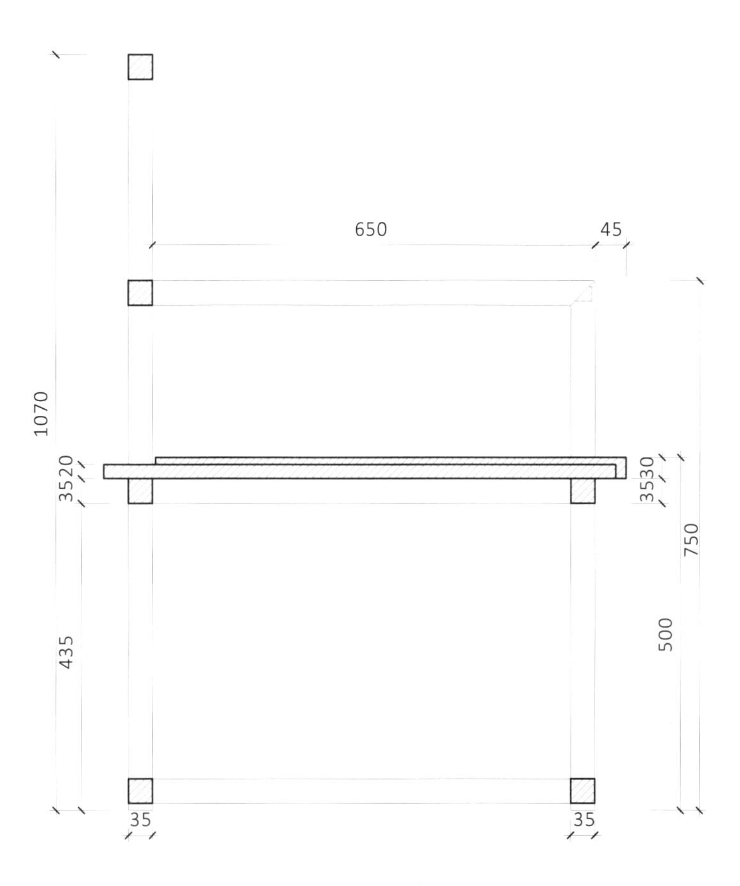

A剖面图 · SIDE AT A ·

刘泽 测绘
LIU ZE DELIN.

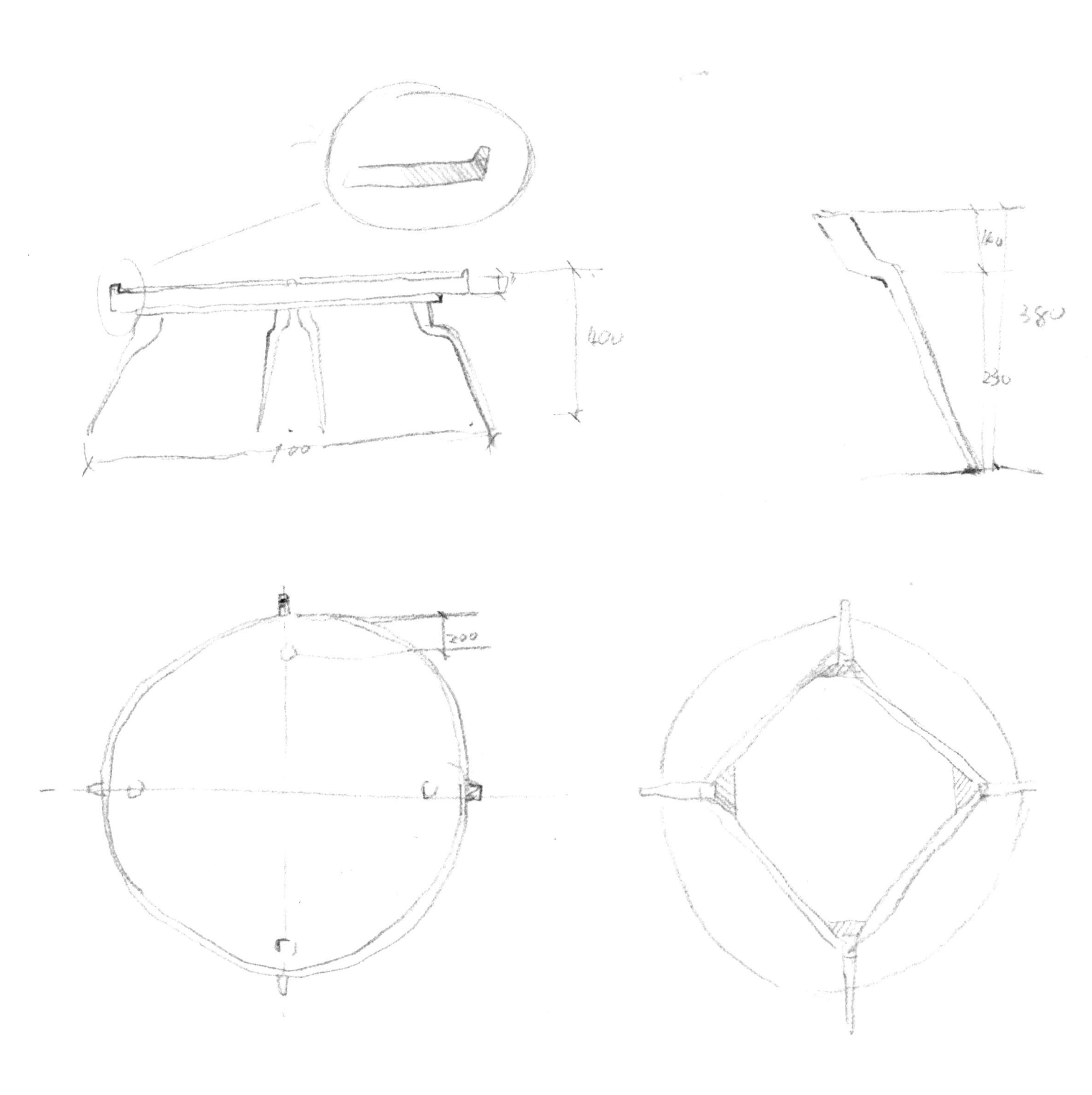

400
140
380
230
200

咖啡几
Coffee table

圆形咖啡几，造型简洁。四条木质桌腿穿过桌面形成4个木质“犄角”，增加了作品的趣味性。

规格 直径 x 高 Ø1050 x 492(mm)
设计者 朱宁克
制作商 Rabiscos Sensatos

It is a round coffee table with simple style. The four wooden legs protrude through the table top, making it fun.

Dimension diameter x height Ø1050 x 492(mm)
Deigner Zhu Ningke
Maker Rabiscos Sensatos

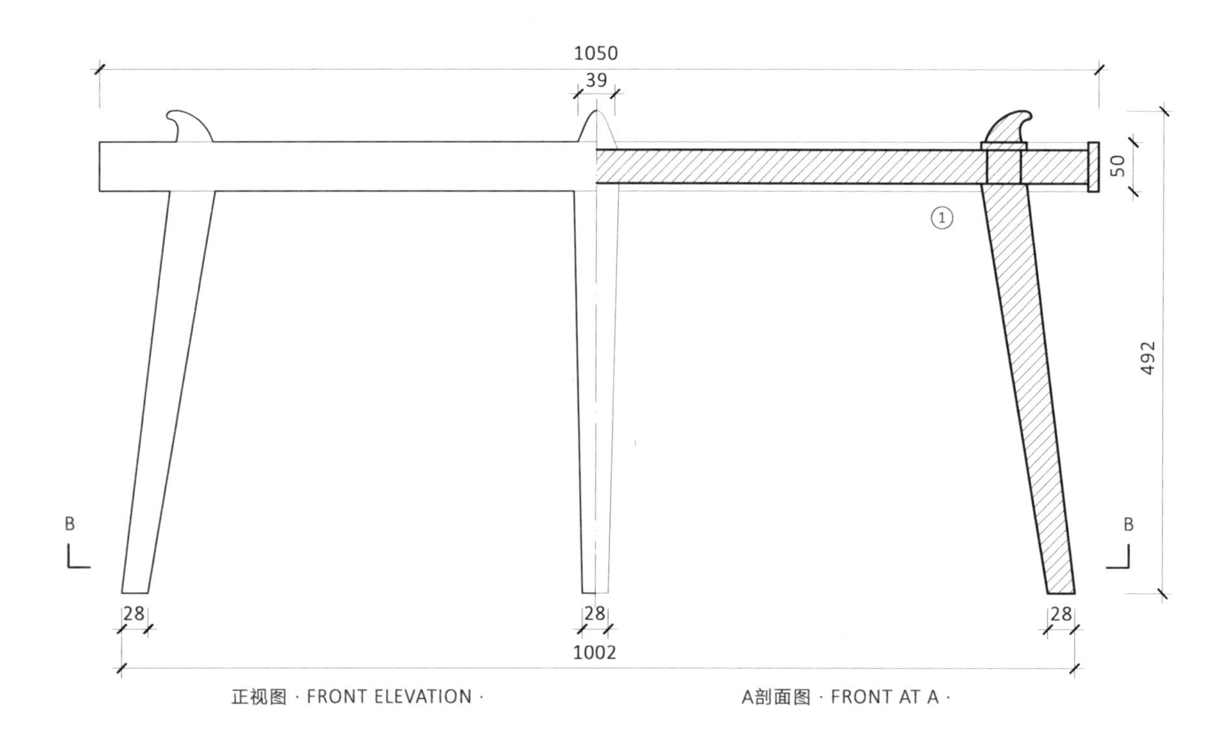

正视图 · FRONT ELEVATION ·

A剖面图 · FRONT AT A ·

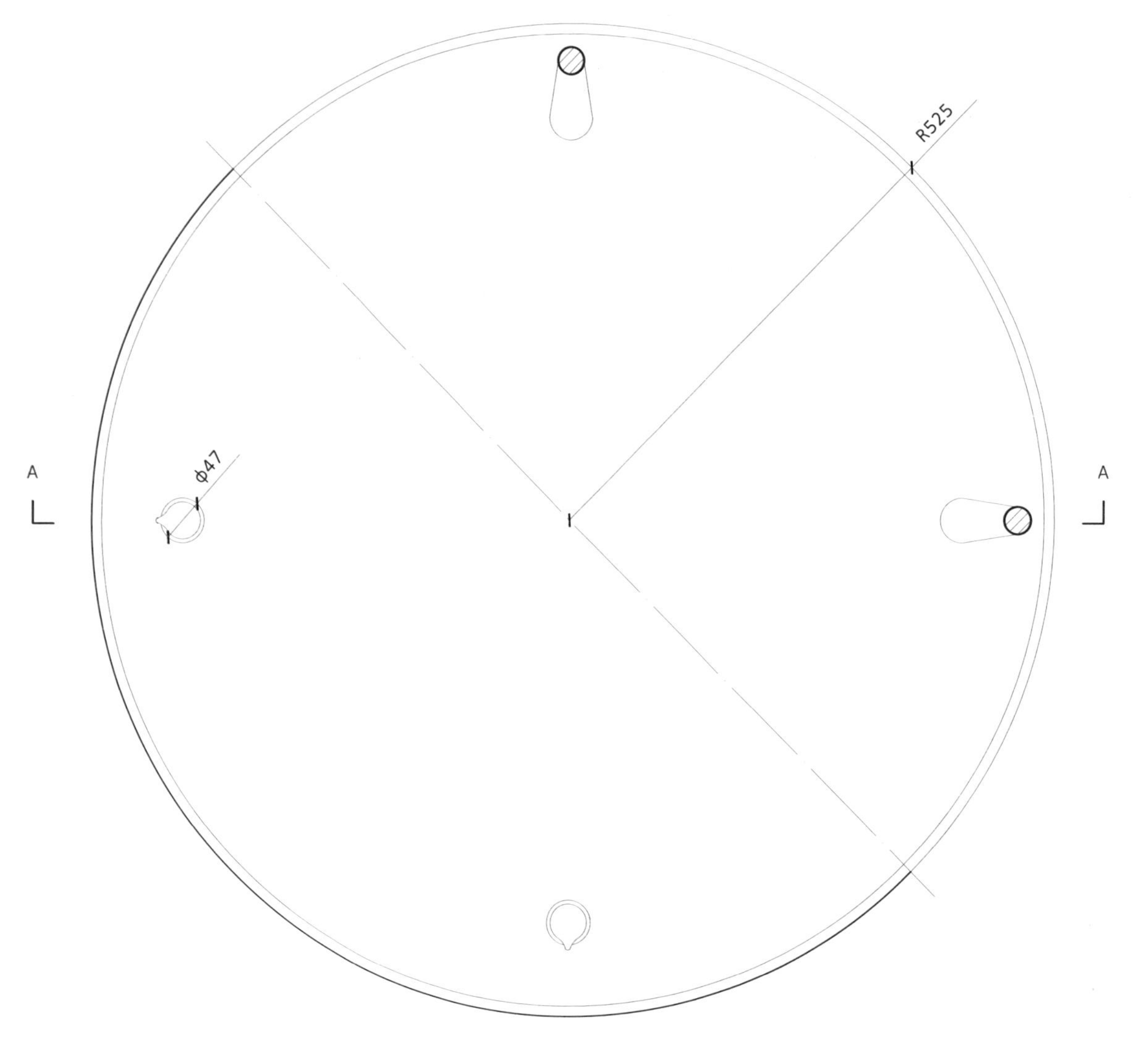

B剖面图 · PLAN AT B ·

顶视图 · PLAN OF TOP ·

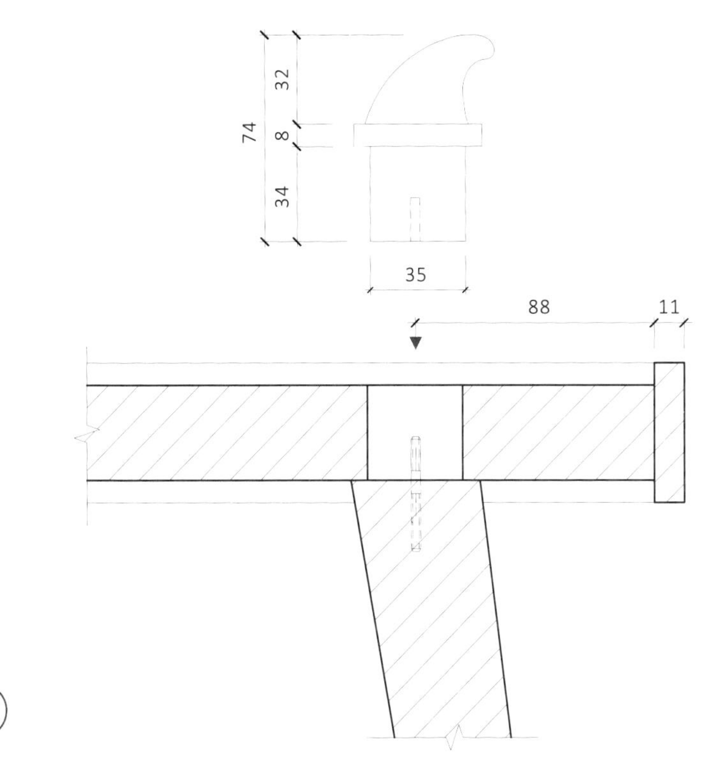

潘阳 测绘
PAN YANG DELIN.

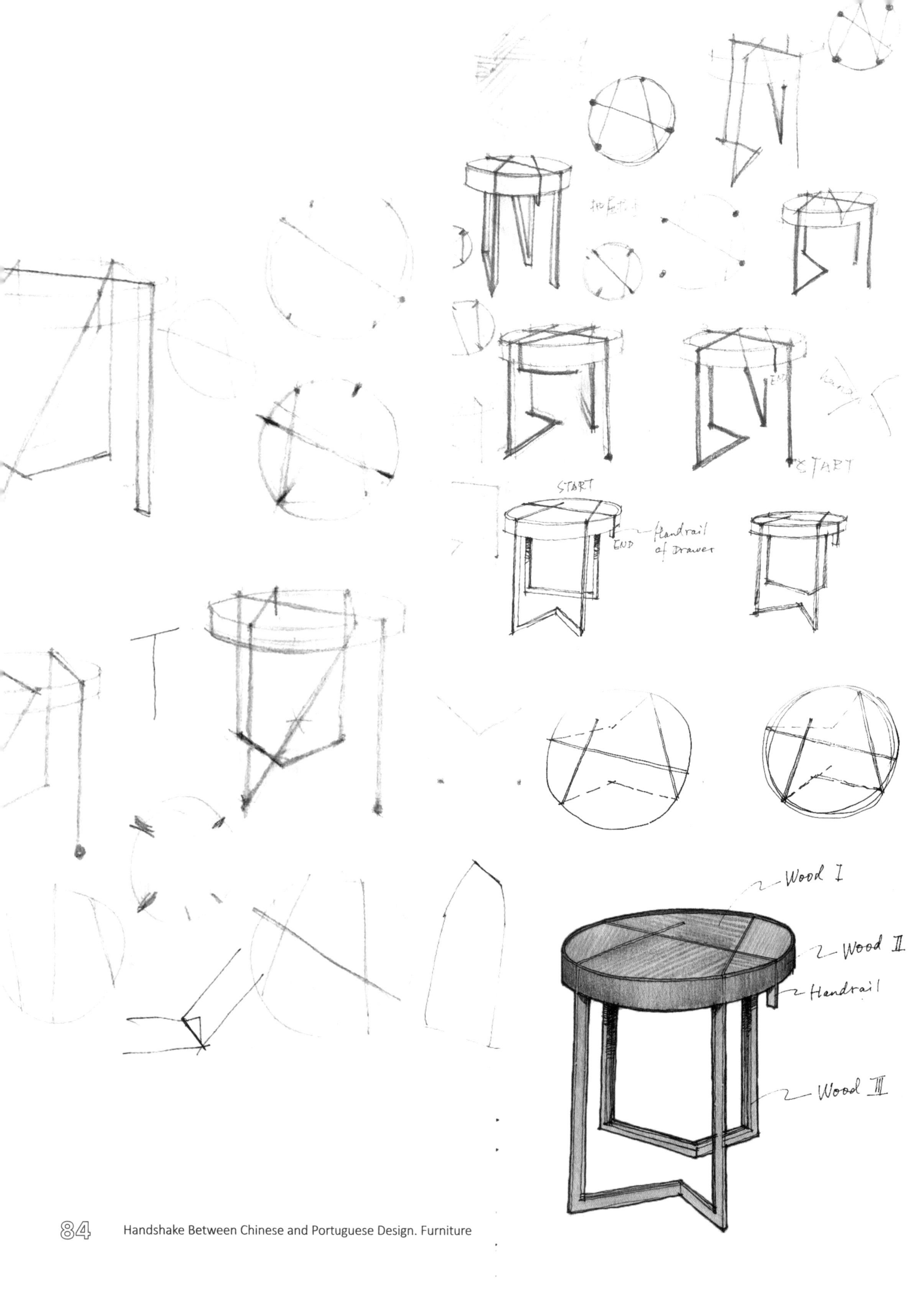
START
END
START
Handrail
END
of Drawer
Wood I
Wood II
Handrail
Wood III

缠绕
Binding

圆几设计围绕日常家用需求展开。以一条缠绕的折线为形态创意，从抽屉拉手开始，与腿足和装饰形成连贯折线，最终止于几面。抽屉内部隔板分布合理，保证日常功能使用。

规格 直径 x 高 Ø650 x 830(mm)
设计者 韩风
制作商 Gualtorres - Design e Indústria de Mobiliário, Lda

The roundtable designs for meeting daily household needs. The inspiration is a binding line starting from the drawer handle, the legs, to the table top. The logical layout of the partition in the drawer makes sound daily use.

Dimension diameter x height Ø650 x 830(mm)
Deigner Han Feng
Maker Gualtorres - Design e Indústria de Mobiliário, Lda

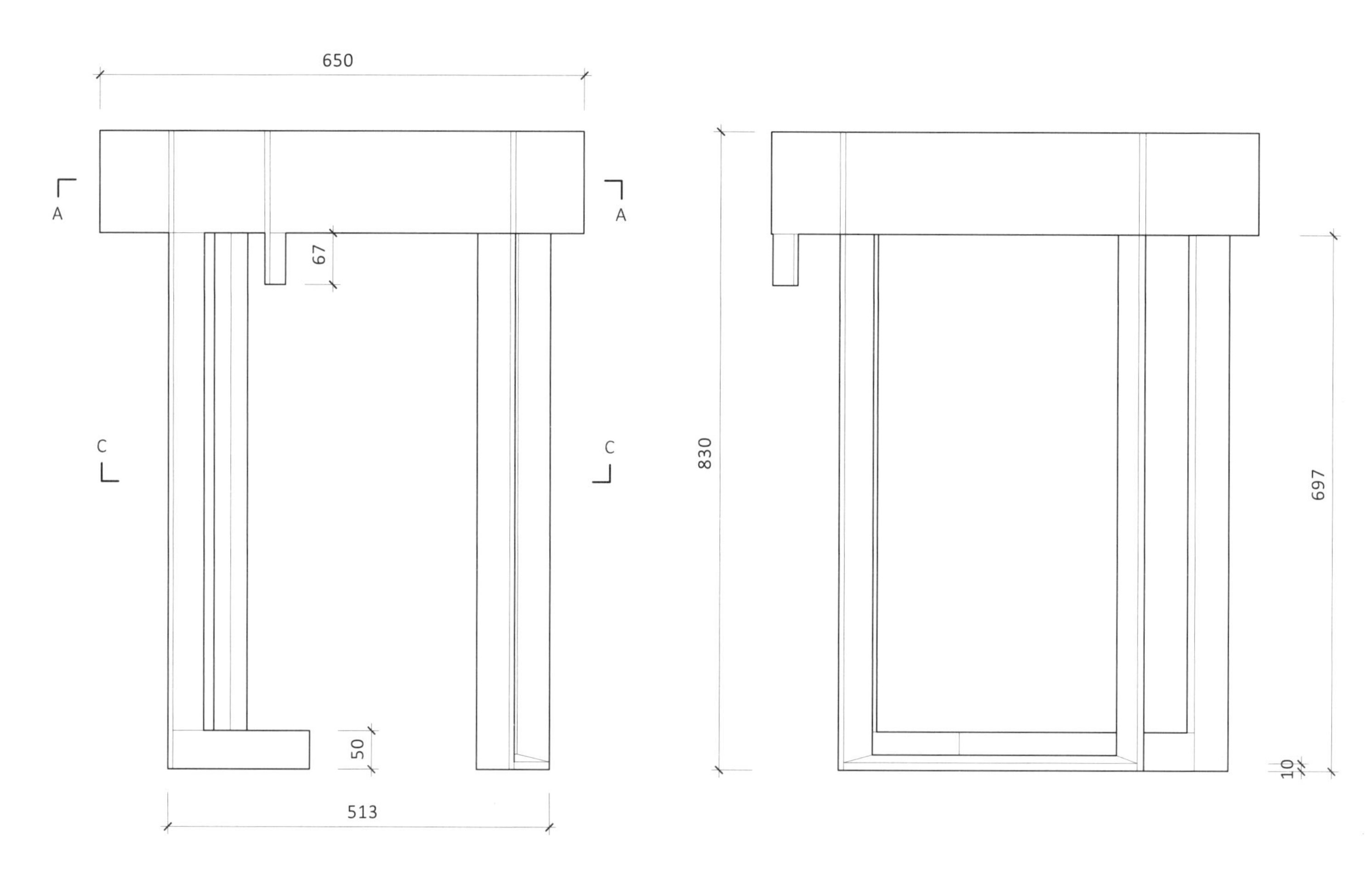

正视图 · FRONT ELEVATION ·

侧视图 · SIDE ELEVATION ·

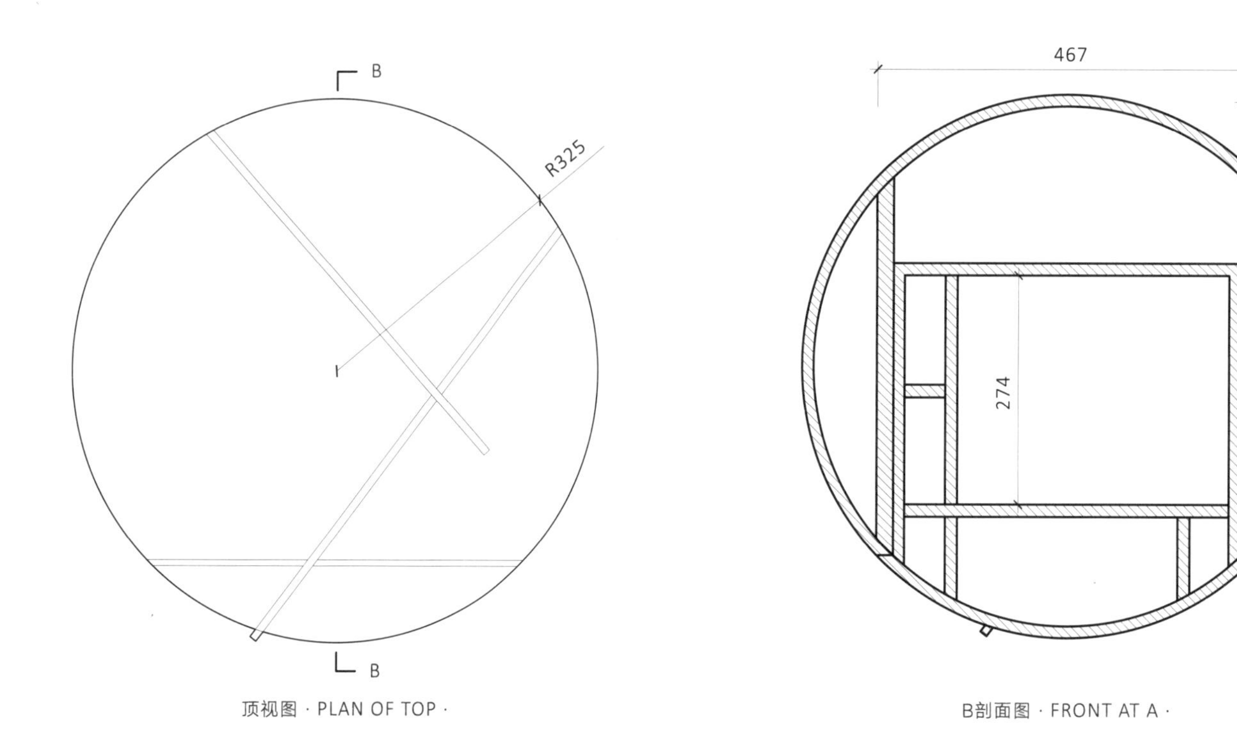

顶视图 · PLAN OF TOP ·

B剖面图 · FRONT AT A ·

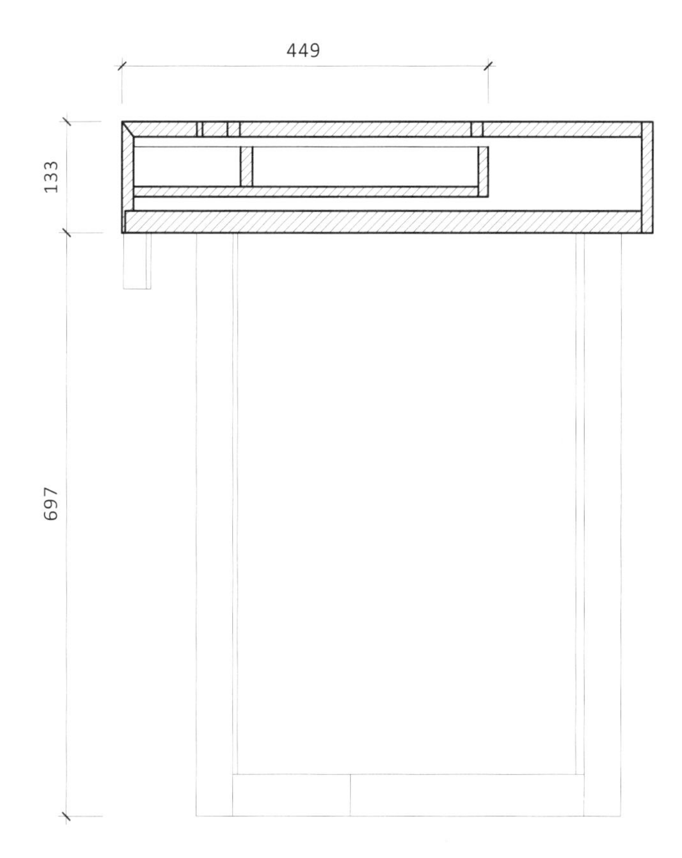

A剖面图 · FRONT AT A ·

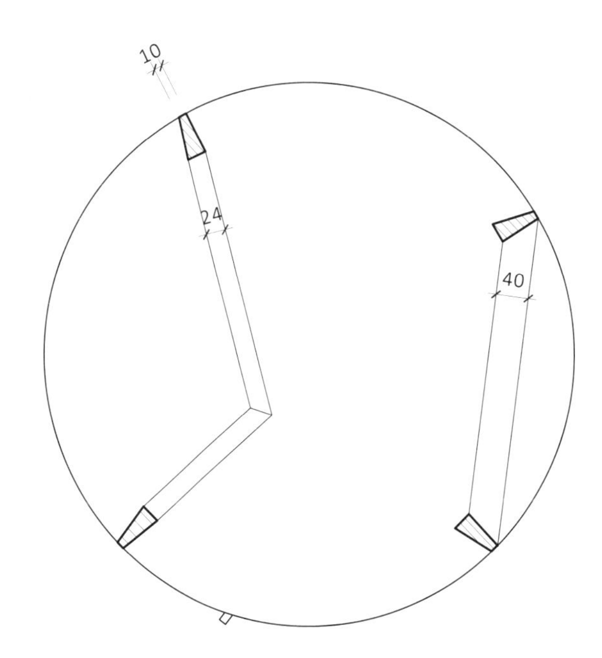

C剖面图 · FRONT AT C ·

潘阳 测绘
PAN YANG DELIN.

妆·纳
Sensation

化妆桌表面看起来像一张开心的笑脸，赋予此作品随性、开朗的性格。造型简洁、占用空间小。揭开桌面是一面镜子，桌体设上下两层，可分层存放彩妆、首饰，适用于年轻女性。

规格 长 x 宽 x 高 547 x 290 x 820(mm)
设计师 孙滨
制作商 Camila Móveis Indústria Mobiliário Lda.

The surface of the dresser is a smiling face, giving a casual and cheerful feeling. The design is space-saving and straightforward. You will see a mirror when uncovering the table top. There are two shelves inside for keeping cosmetics and jewelry.

Dimension length x width x height 547 x 290 x 820(mm)
Deigner Sun Bin
Maker Camila Móveis Indústria Mobiliário Lda.

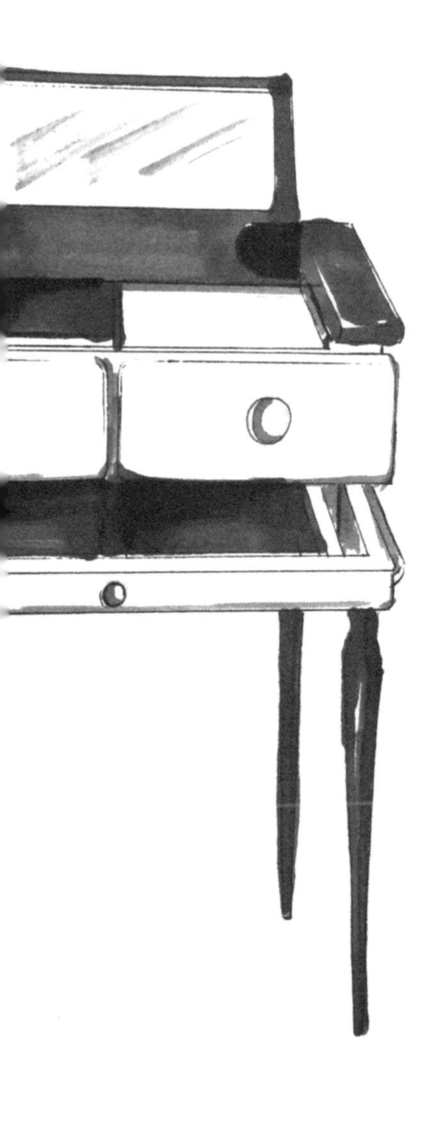

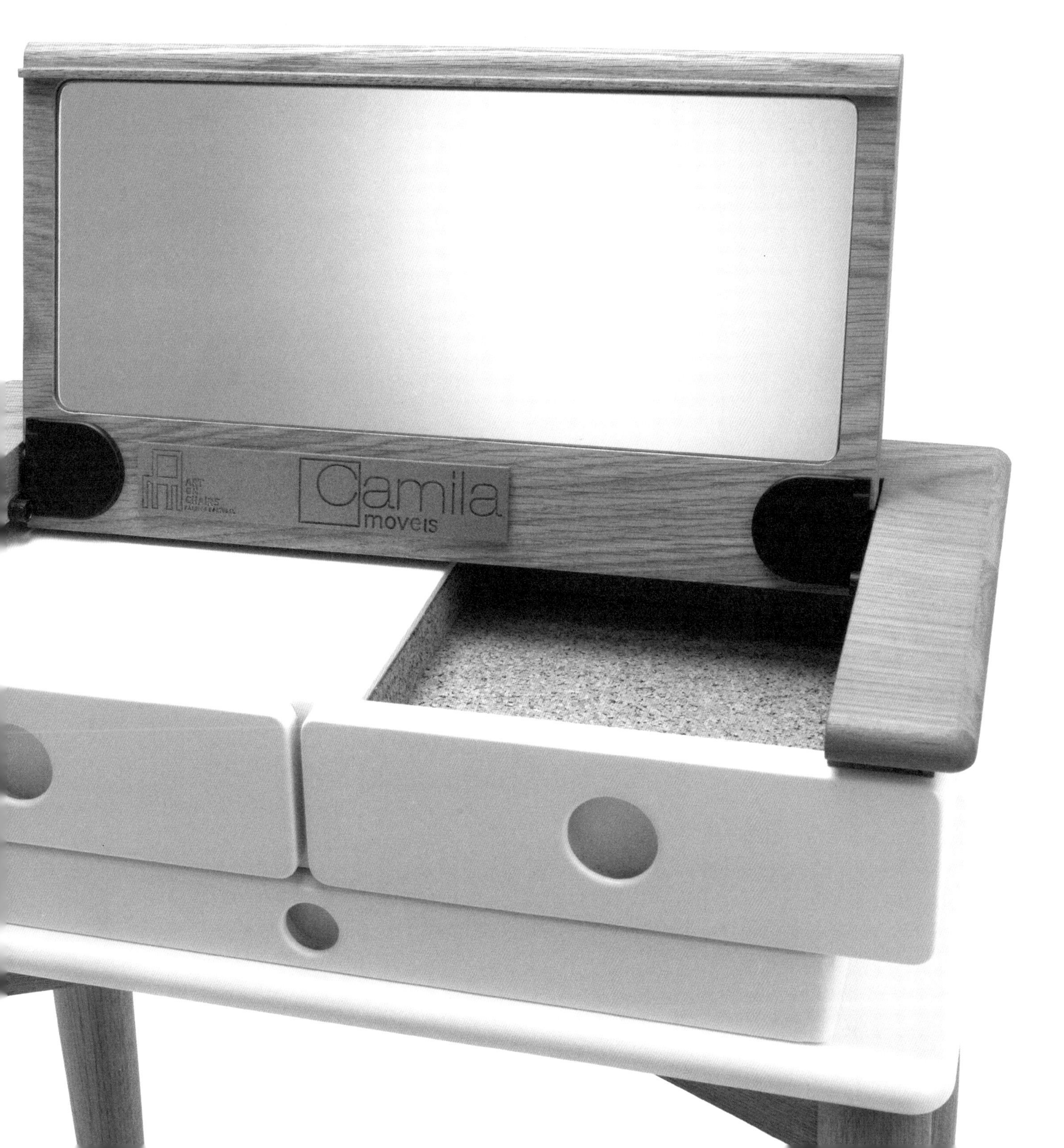
Camila
moveis

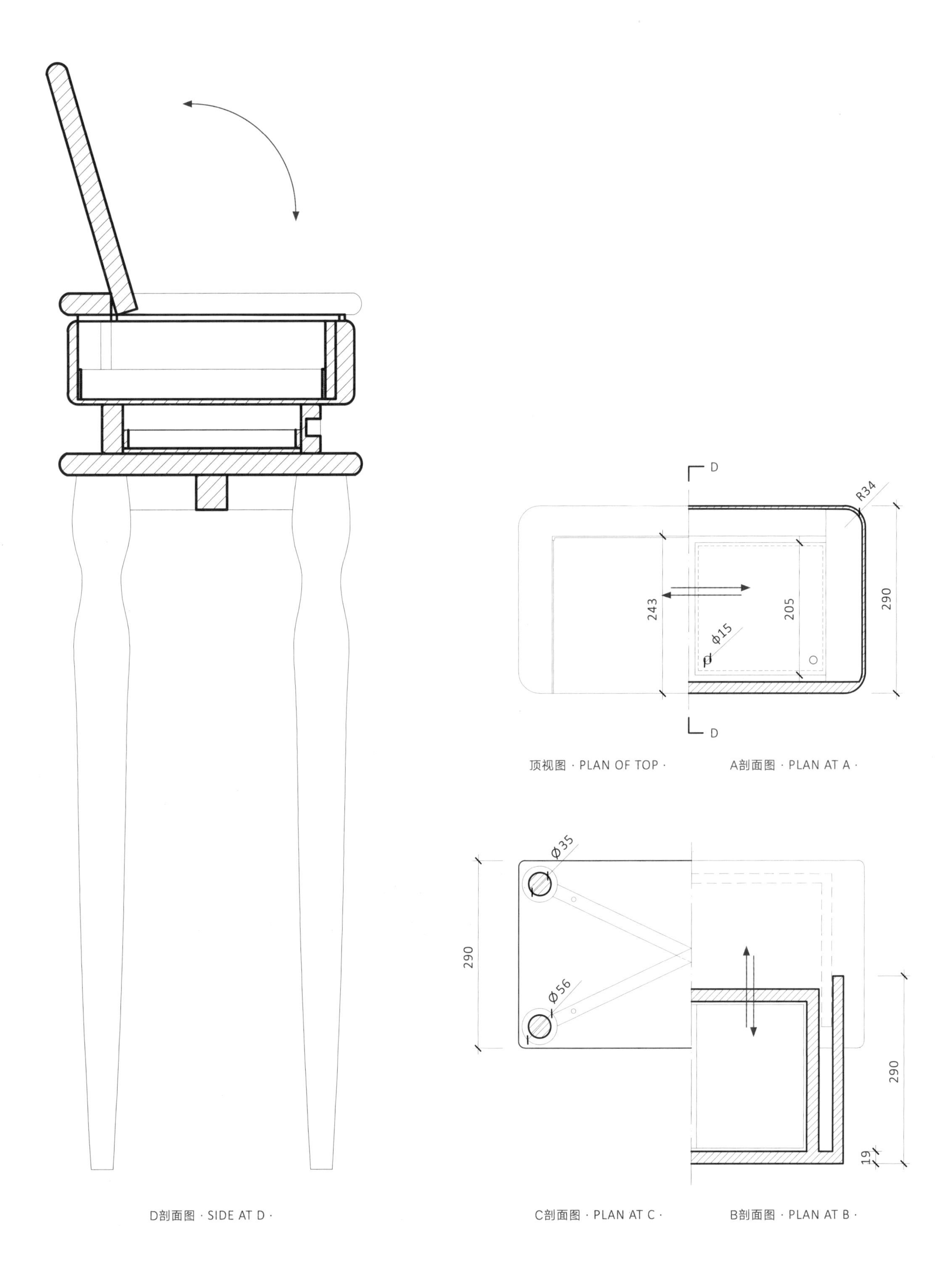

D剖面图 · SIDE AT D ·

顶视图 · PLAN OF TOP ·

A剖面图 · PLAN AT A ·

C剖面图 · PLAN AT C ·

B剖面图 · PLAN AT B ·

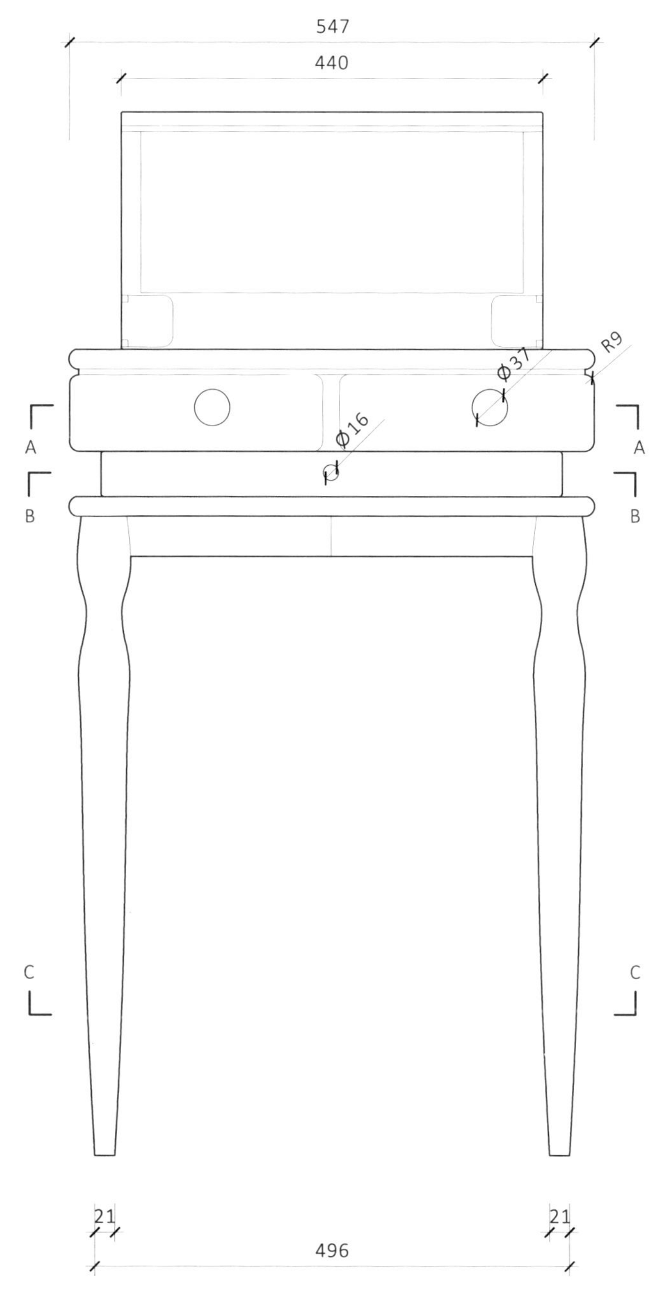

正视图 · FRONT ELEVATION ·

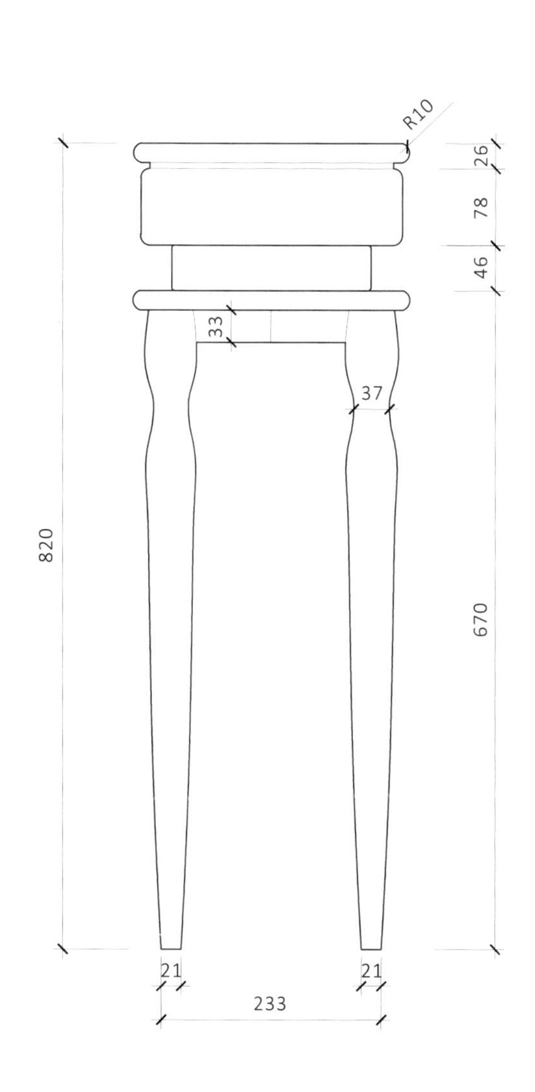

侧视图 · SIDE ELEVATION ·

孙小鹏 测绘
SUN XIAOPENG DELIN.

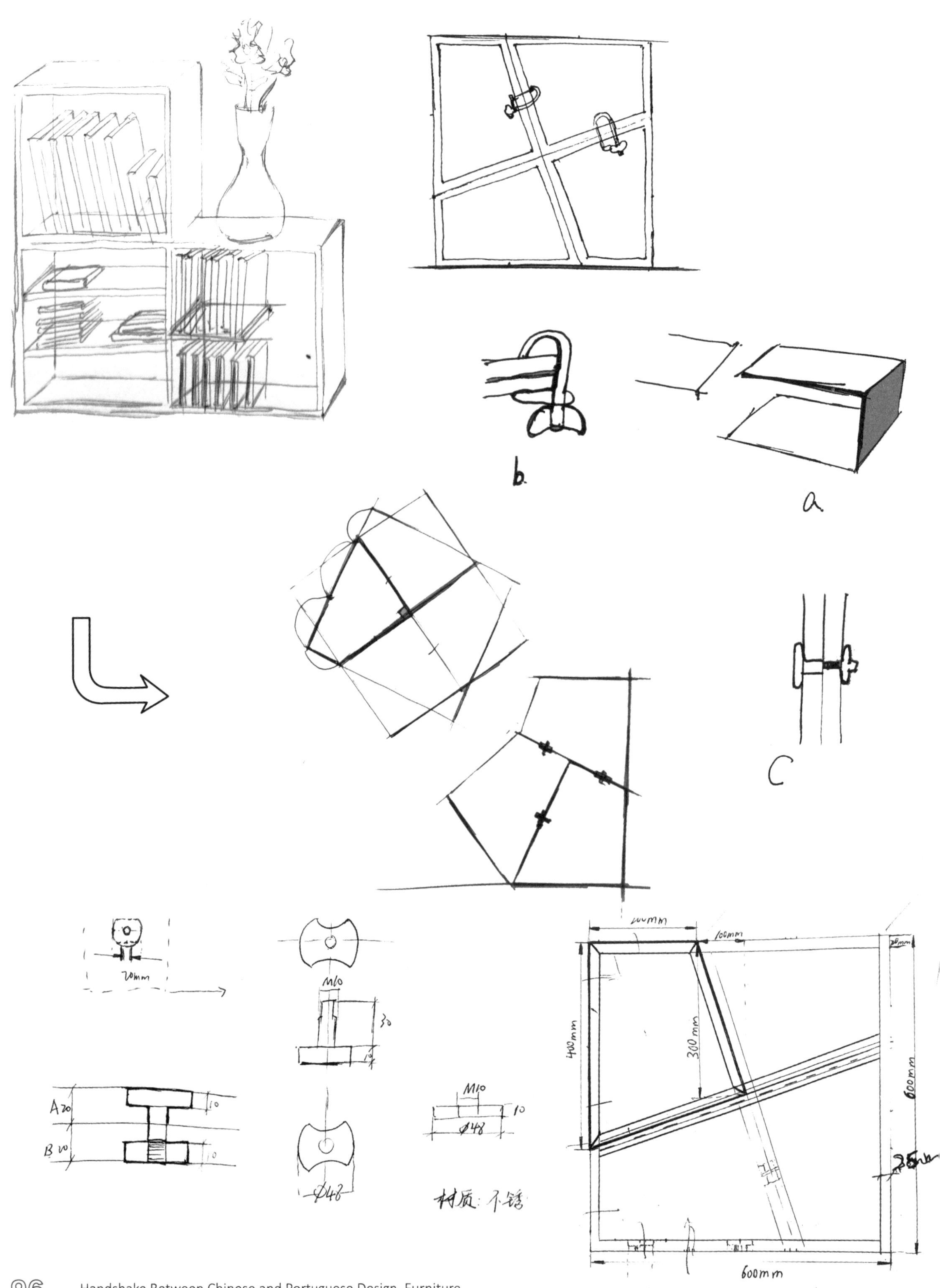
b
a
C
20mm
M10
30
10
A20
B10
10
10
M10
10
Ø48
Ø48
材质: 不锈
400mm
100mm
20mm
400mm
300mm
600mm
25mm
600mm

分・格空间
Separate space

此置物架的设计思路寻求个性化需求与批量生产相结合，以适应不同空间环境为目的。在相同单元的不同组合中，人们融入自己的创意，在规则与自由、理性与感性之中探索独特的表达方式。

规格（正方组合单元） 长 x 宽 x 高 600 x 300 x 600(mm)
设计师 闫卓远
制作者 ABrito mobiliário

The shelf pursues both individual needs and mass production and aims to fit different space. Through different combination, users can add their creativity to the same area, thus achieving unique expression between rules and freedom, sense and sensibility.

Dimension(Square Unit) length x width x height 600 x 300 x 600(mm)
Deigner Yan Zhuoyuan
Maker ABrito mobiliário

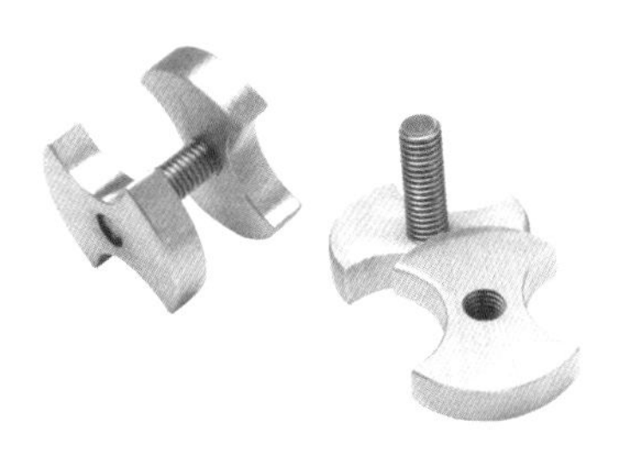

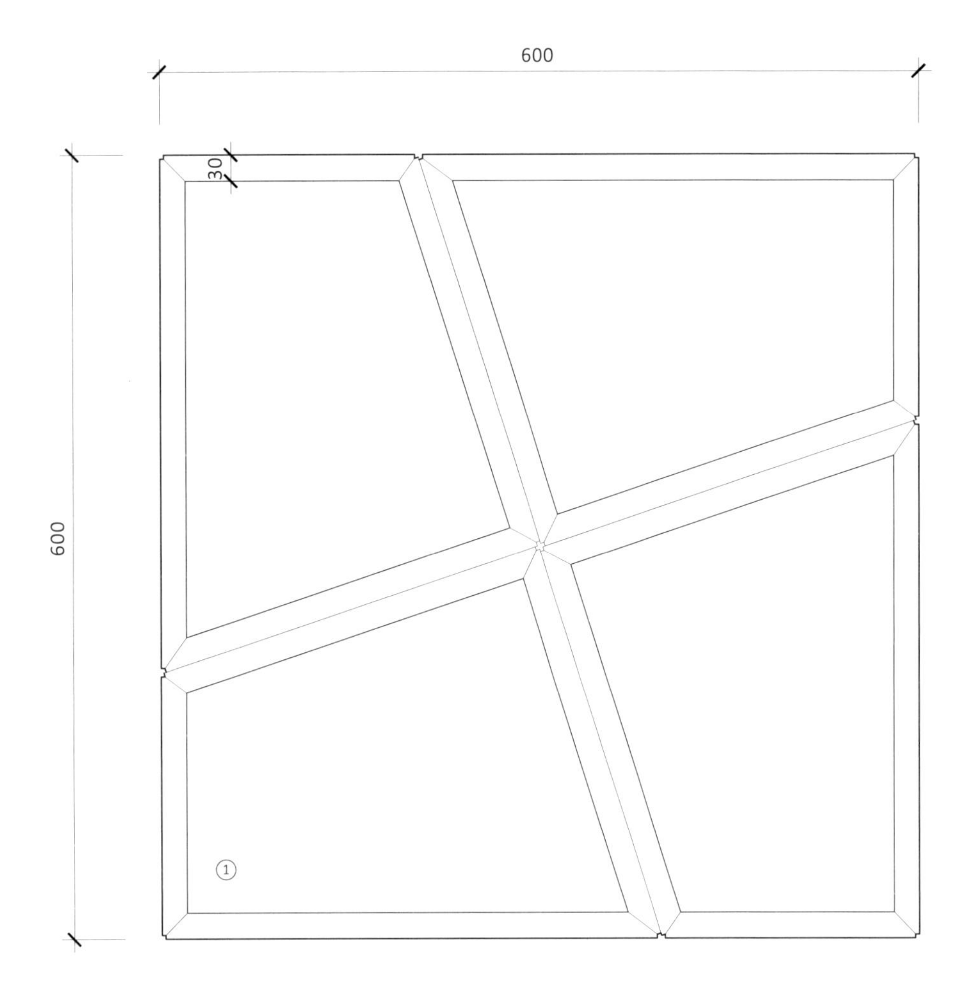

正视图 · FRONT ELEVATION ·

300
A
A

侧视图 · SIDE ELEVATION ·

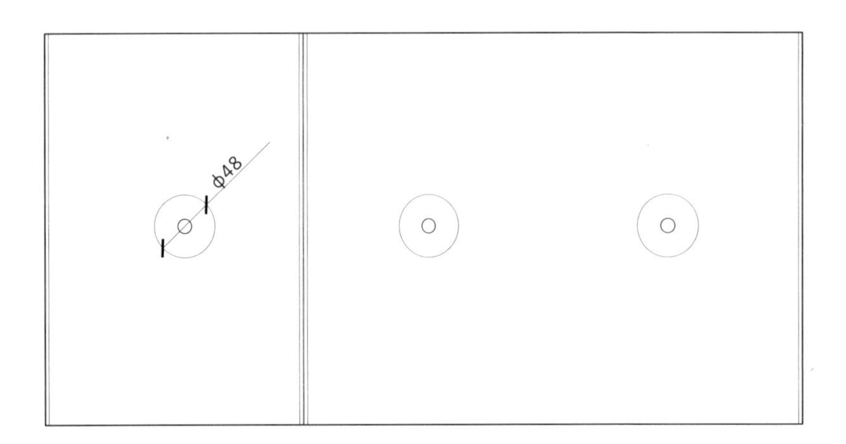

顶视图 · PLAN OF TOP ·

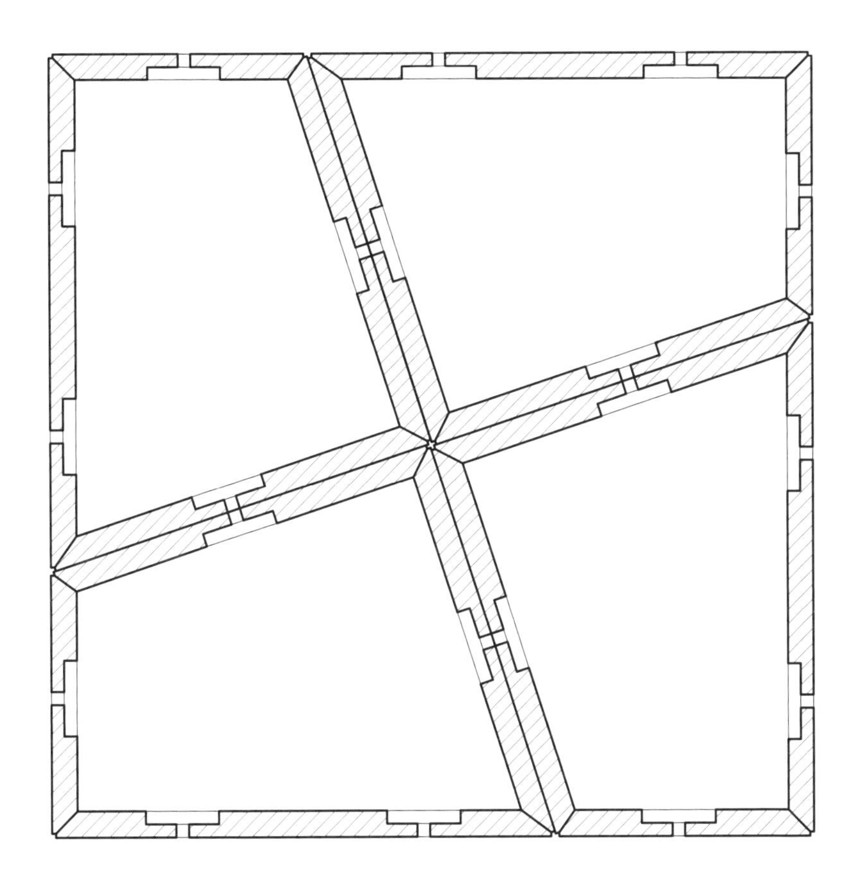

A剖面图 · PLAN AT A ·

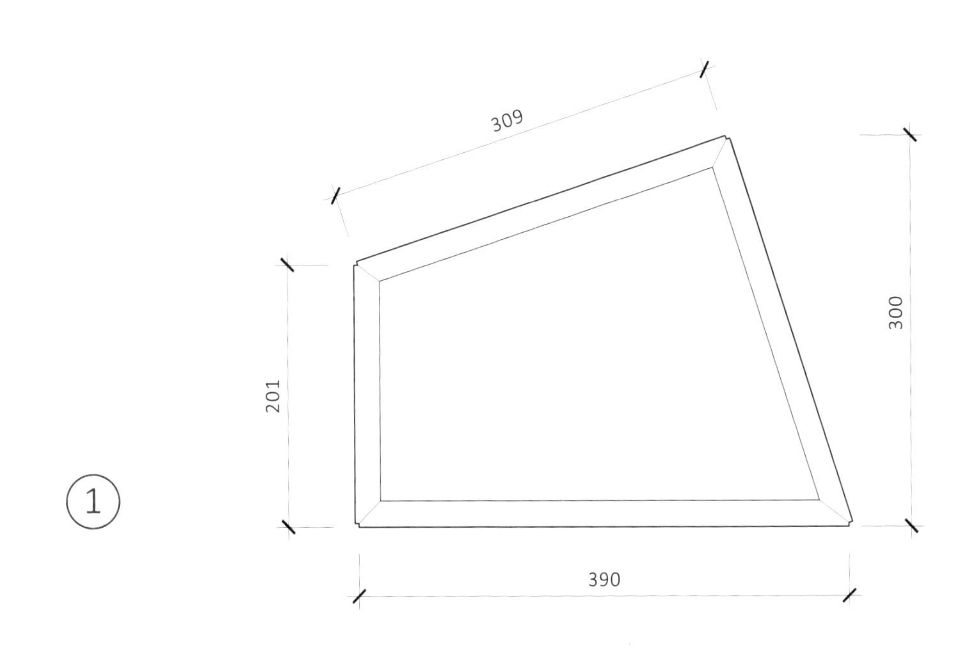

孙小鹏 测绘
SUN XIAOPENG DELIN.

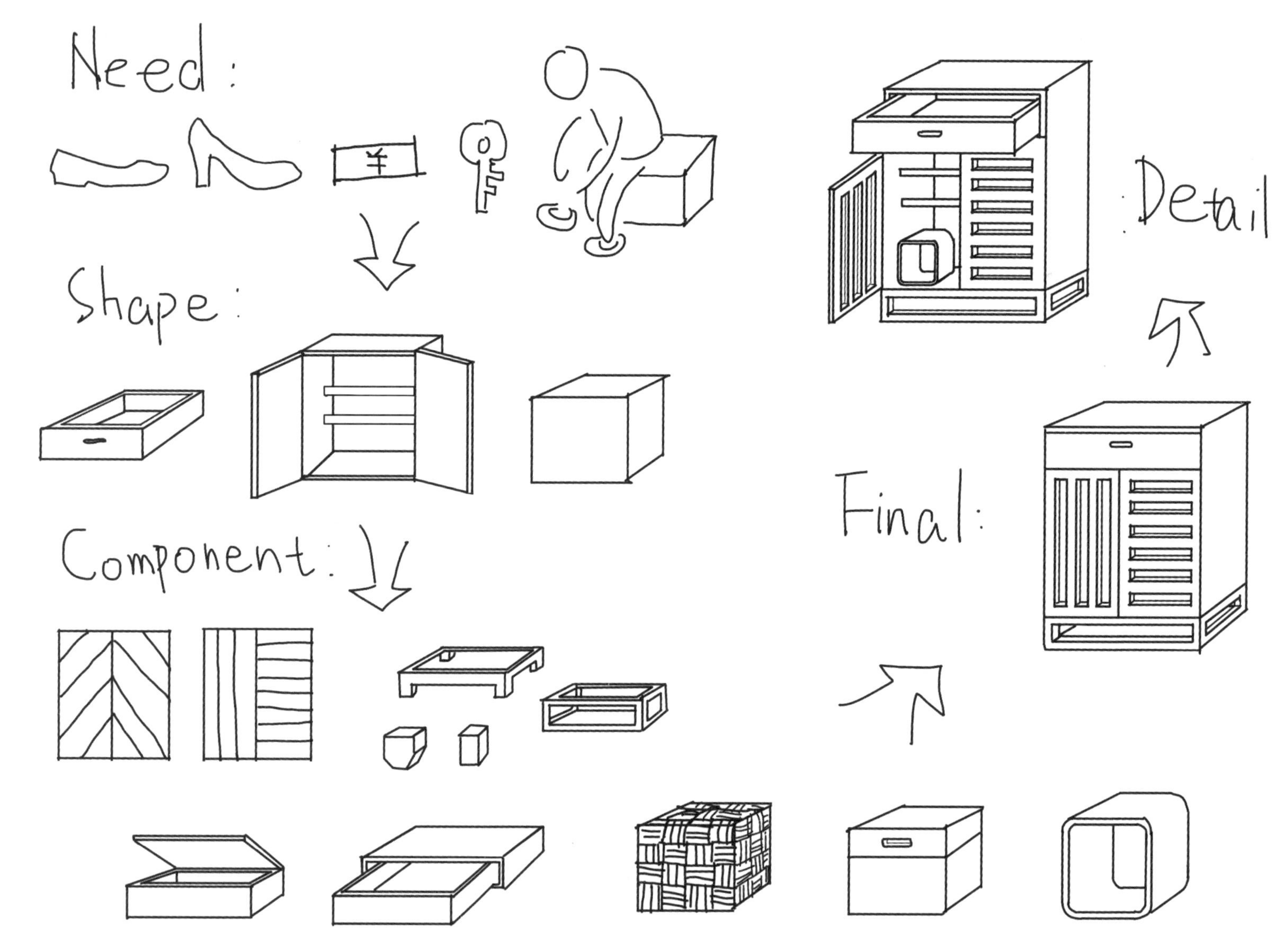
Need:
Shape:
Component:
Detail
Final:

清风
Breeze

作品灵感来源于栏栅，横纵结合,穿插有序，通过最简约的直线条构成产品外观和结构。主要用作鞋柜，上层抽屉可存放小杂物，镂空门扇便于通风，配以鞋凳方便使用。

规格（鞋柜） 长 x 宽 x 高 906 x 320 x 1115(mm)
规格（小凳） 长 x 宽 x 高 300 x 258 x 300(mm)
设计者 赵安路
制作商 Comodos Mobiliário, Lda.

Fences inspire the design. The appearance and structure of this work formed with elegant lines. It used as a shoe cabinet. Its upper drawer is for keeping small things and the hollow-out door for ventilation. It would be more convenient when used together with a stool.

Dimension(shoe cabinet) length x width x height 906 x 320 x 1115(mm)
Dimension(stool) length x width x height 300 x 258 x 300(mm)
Deigner Zhao Anlu
Maker Comodos Mobiliário, Lda.

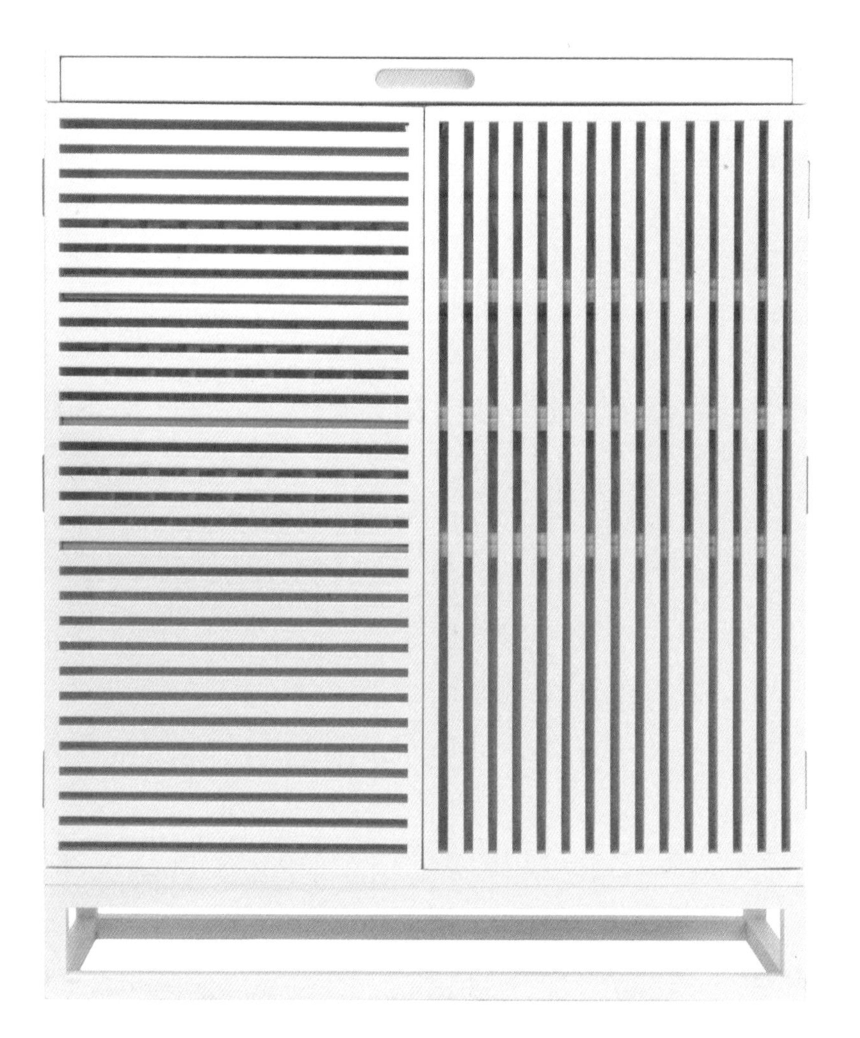

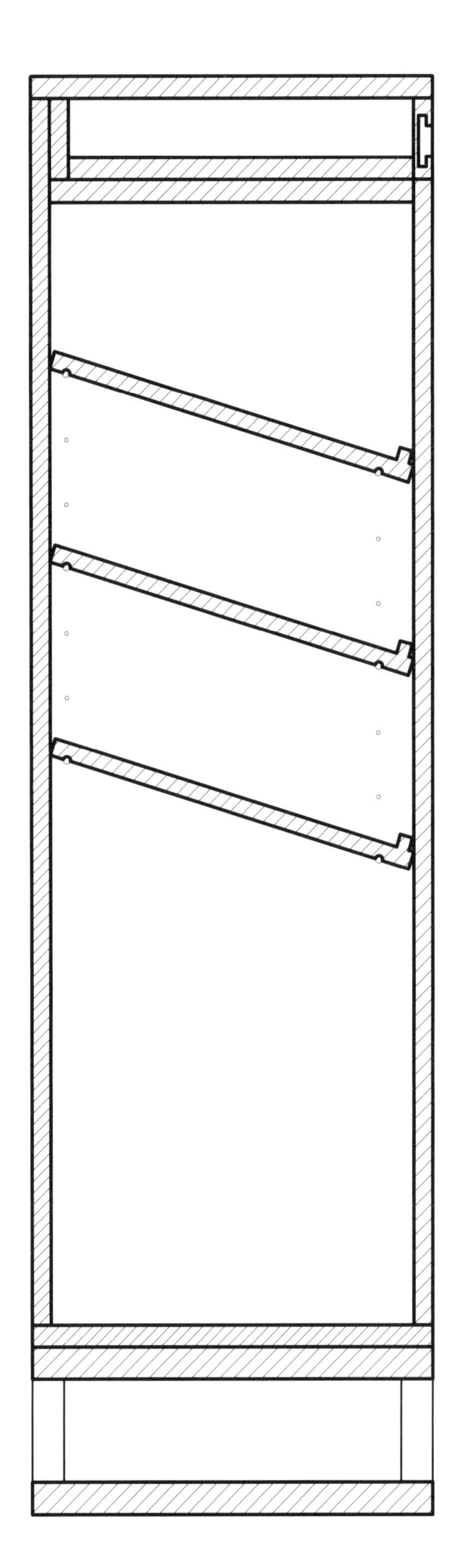

B剖面图 · SIDE AT B ·

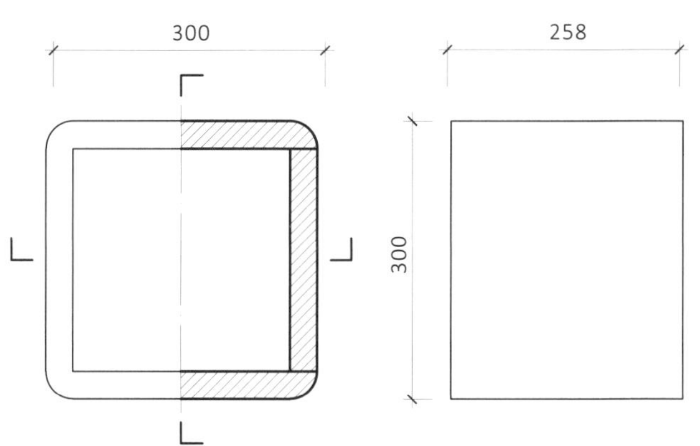

小凳主视图 · STOOL FRONT ELEVATION · 小凳侧视图 · STOOL SIDE ELEVATION ·

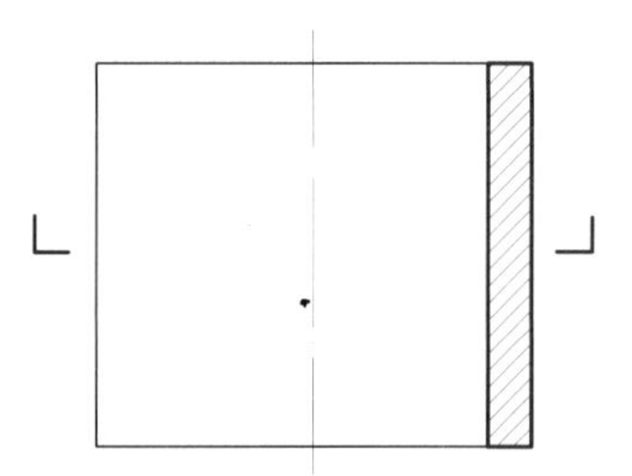

小凳顶视图 · STOOL PLAN ELEVATION ·

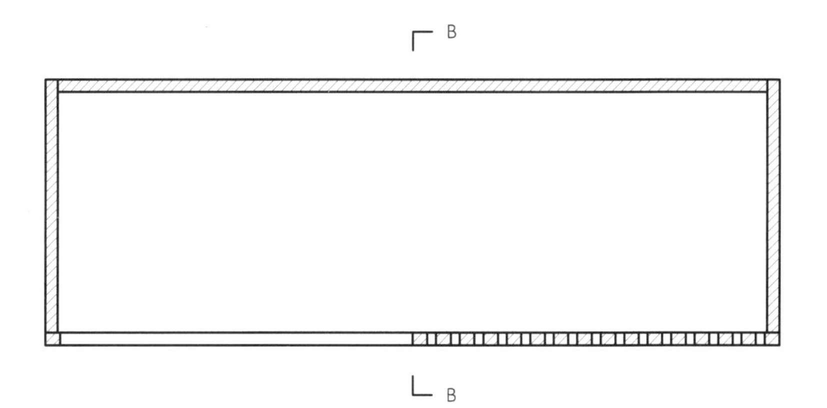

A剖面图 · PLAN AT A ·

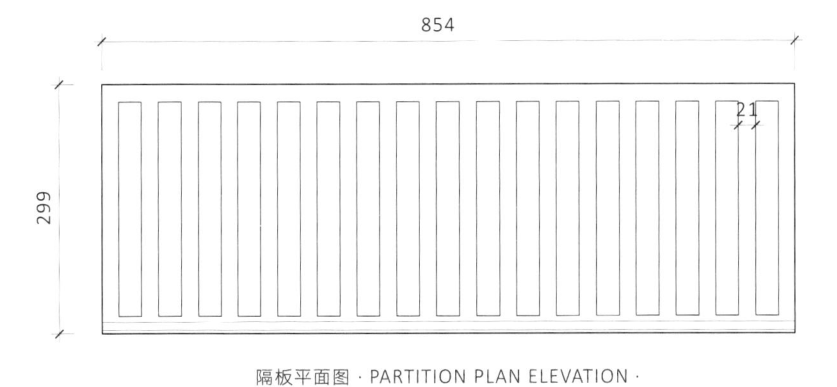

隔板平面图 · PARTITION PLAN ELEVATION ·

906
870
120
R12
18
A
A
1115
25
25
25

正视图 · FRONT ELEVATION ·

侧视图 · SIDE ELEVATION ·

李净尘 测绘
LI JINGCHEN DELIN.

底座：防止卫生间内水汽变形，方便打扫 h/10cm 粘上脚

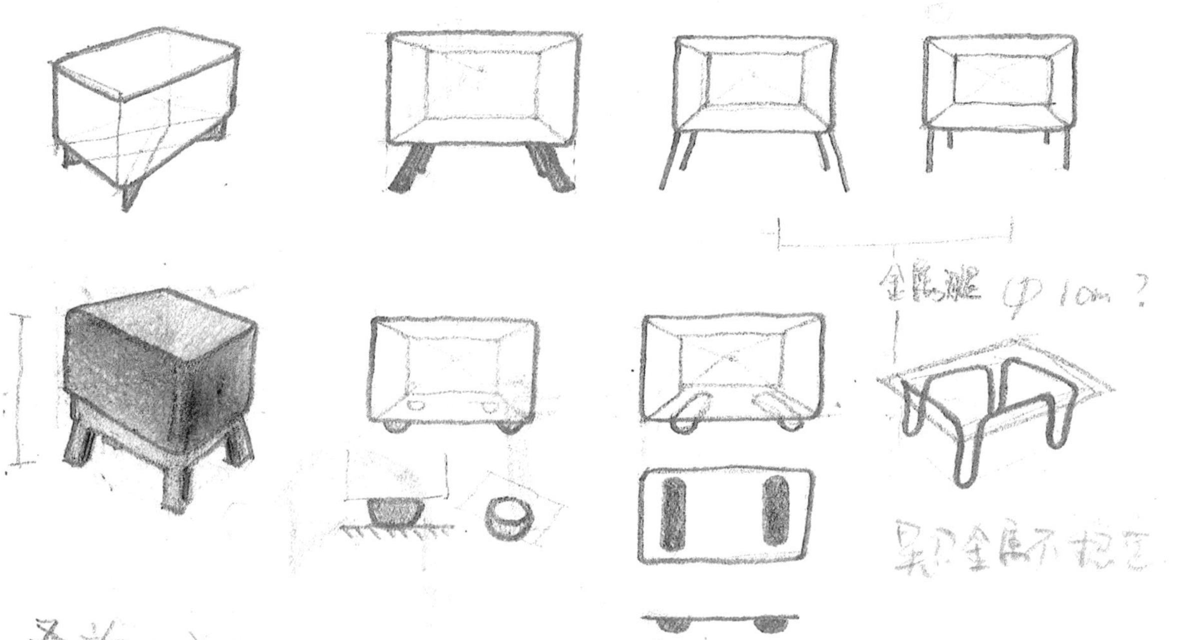

金属腿 φ1cm？

只有金属不稳定

叠放：会误用成小板凳吗？

总h：35cm 小红尺：25×25×25

板凳高度：10

小板凳高度：5cm

25

25

10

cm

布？皮？

布套

+

木壳&板凳

5 25 5

25

20

2

25

cm

割面

FAIA

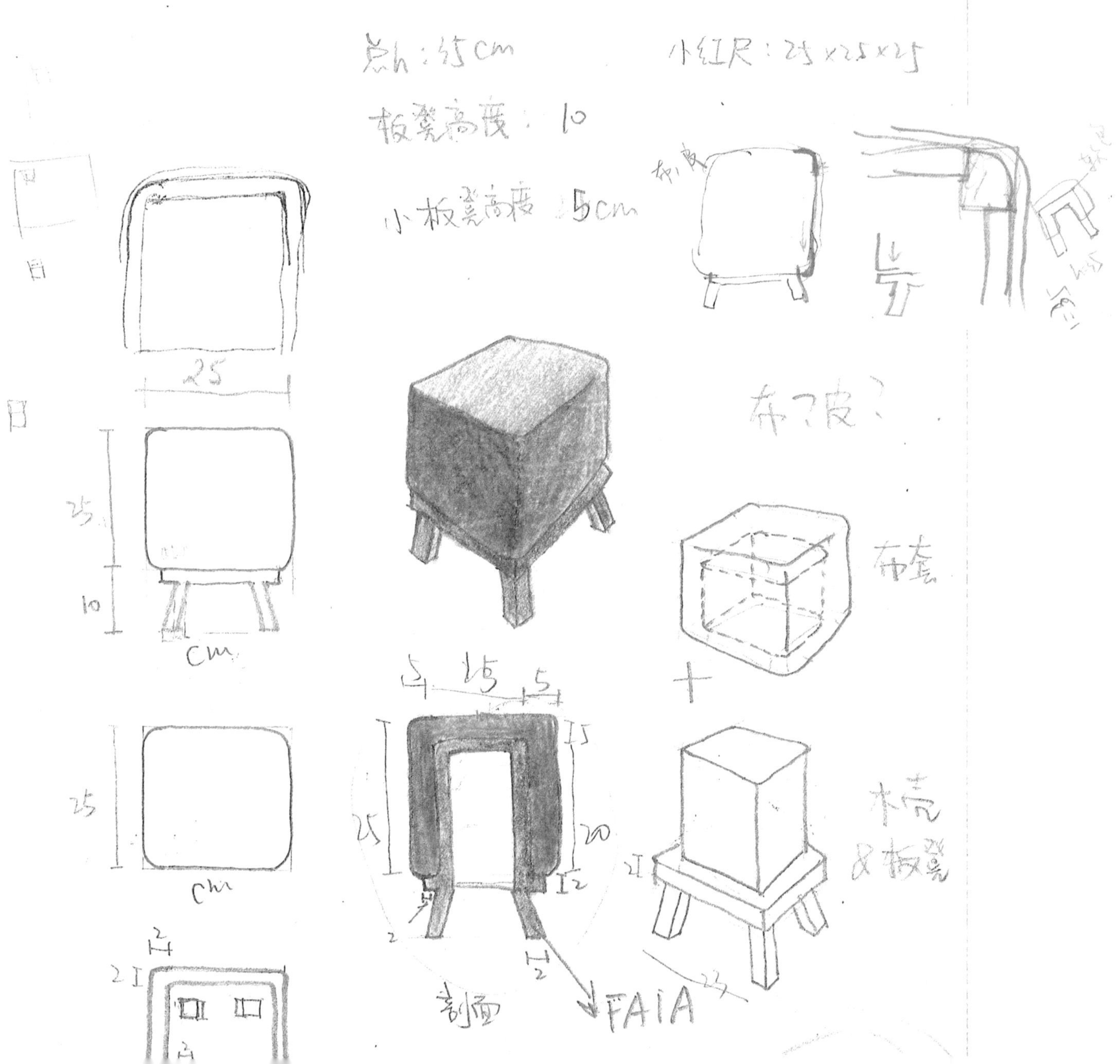

一方凳
A cube

坐凳设计灵感来源于一块积木。色彩鲜亮的小方块凸显出柔软舒适的坐面，底部支撑结构则增强了座具的稳定性。

规格 长 x 宽 x 高 263 x 263 x 353(mm)
设计者 沈相宜
制作商 Comodos Mobiliário, Lda.

The inspiration of a cube was come from a piece of woodblock. The bright-colored block highlights the comfort and softness of seat and the bottom provides stable support.

Dimension length x width x height 263 x 263 x 353(mm)
Deigner Shen Xiangyi
Maker Comodos Mobiliário, Lda.

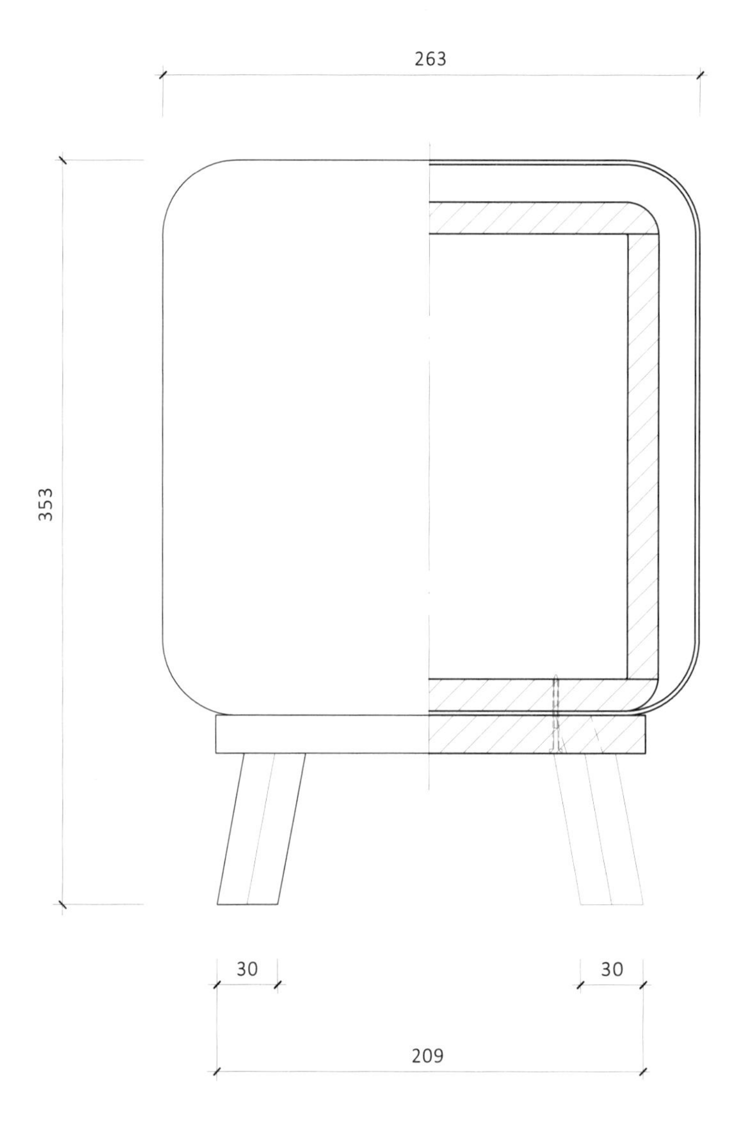

正视图 · FRONT ELEVATION ·

A剖面图 · SIDE AT A ·

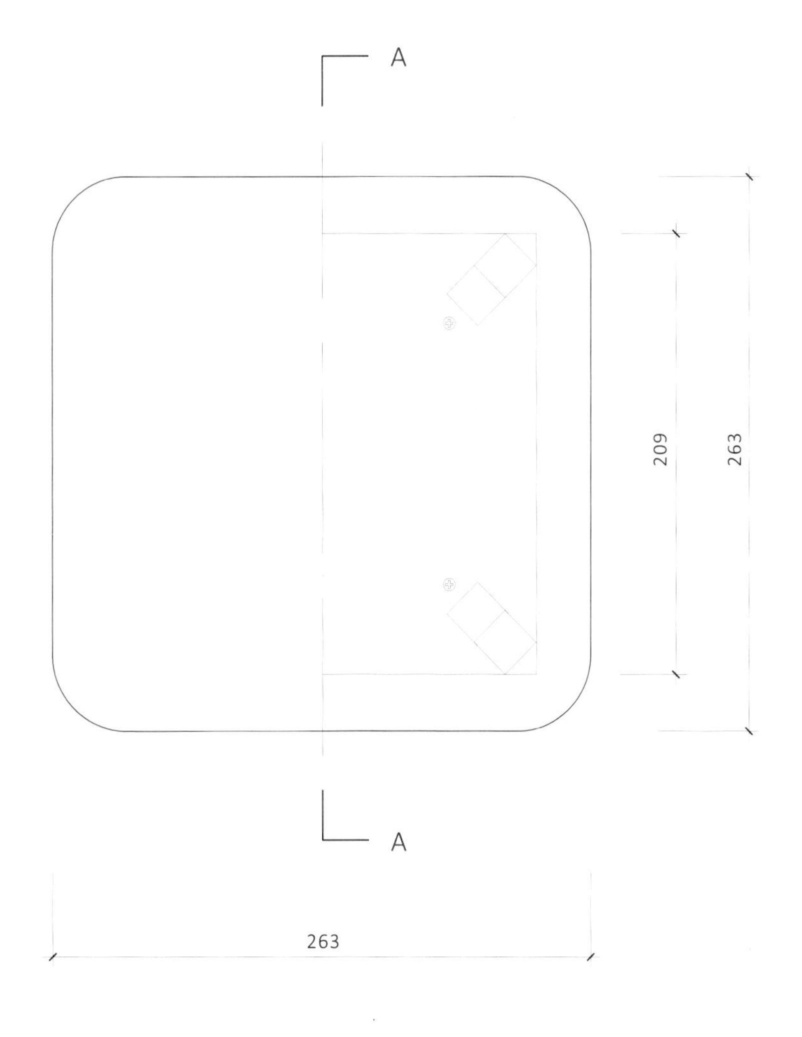

顶视图 · PLAN OF TOP · 底视图 · PLAN LOOKING UP ·

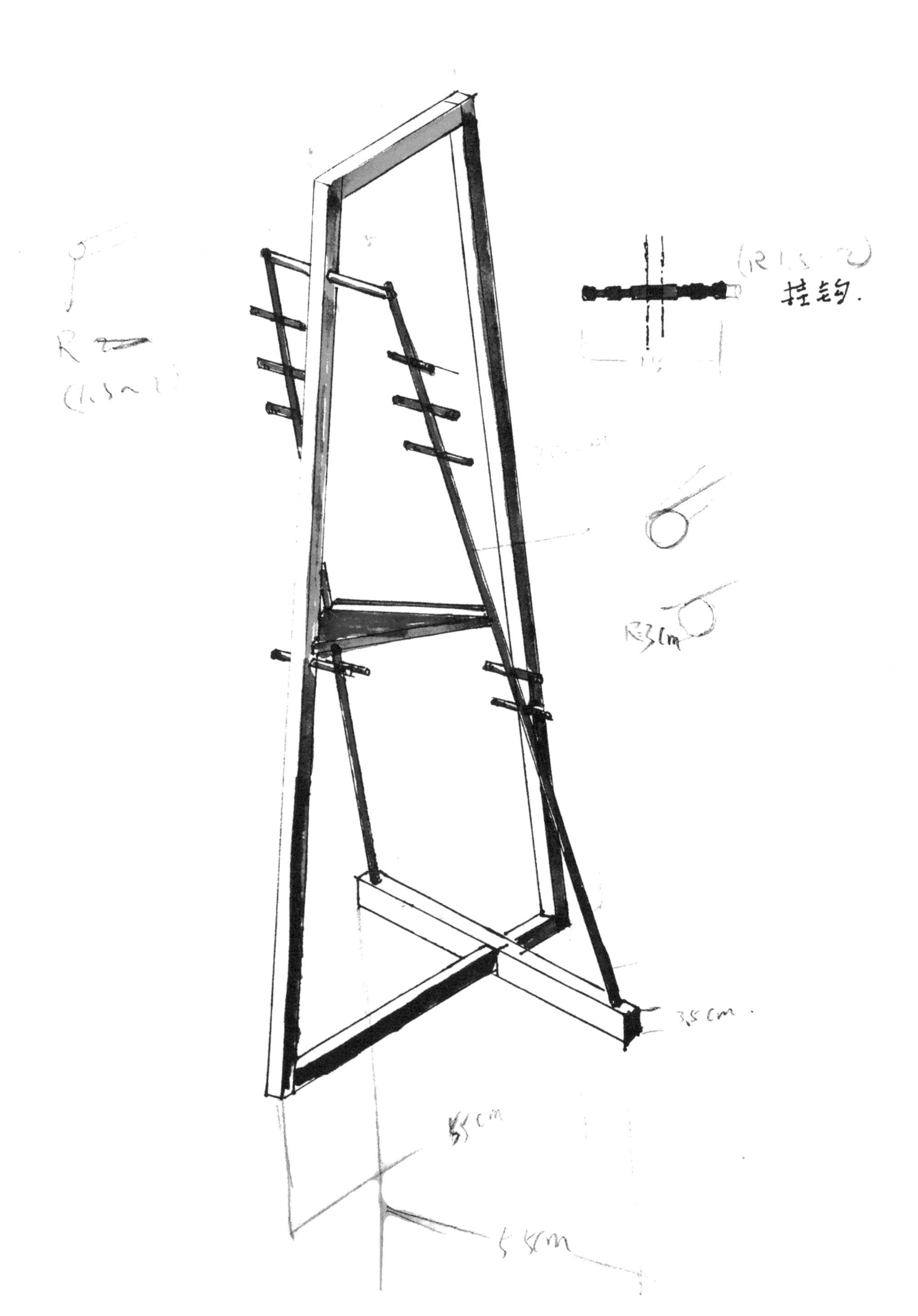
挂钩.
3.5cm

融错·空间
The interlaced dimensions

衣架结构清晰，一方一圆，一纵一横，两者交错，空间融合，使用便捷，是忙碌的城市生活中的一个简单的角落。

规格 长 x 宽 x 高 442 x 447 x 1998(mm)
设计者 赵雄韬
制作商 Rabiscos Sensatos

The clothes stand has the precise structure. It keeps a balance between square and round, vertical and horizontal elements. It provides a user-friendly space in busy city life.

Dimension length x width x height 442 x 447 x 1998(mm)
Deigner Zhao Xiongtao
Maker Rabiscos Sensatos

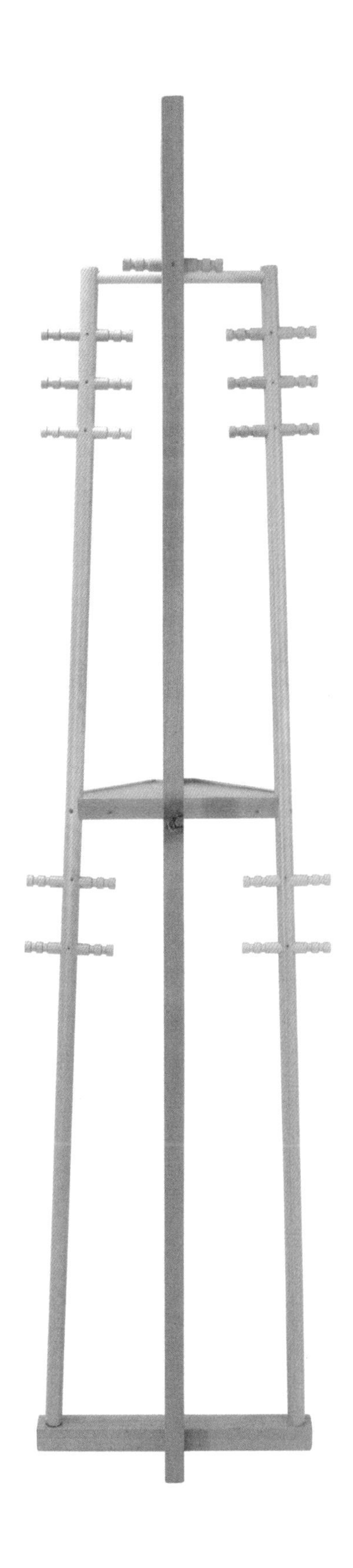

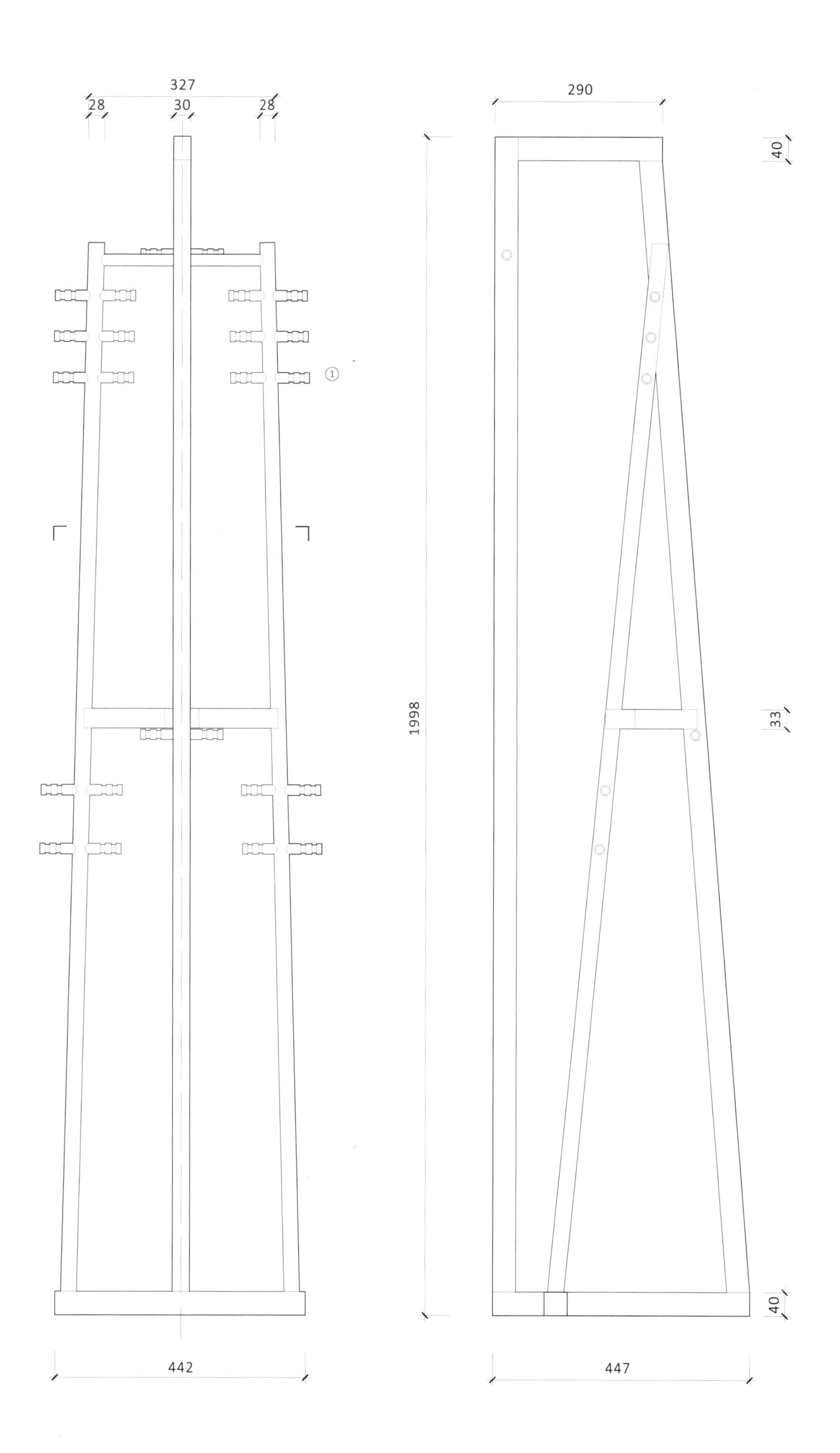
327
28
30
28
290
40
1
1998
33
40
442
447

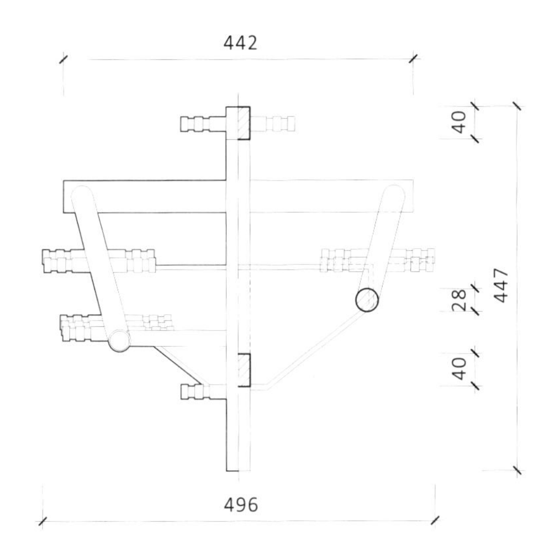

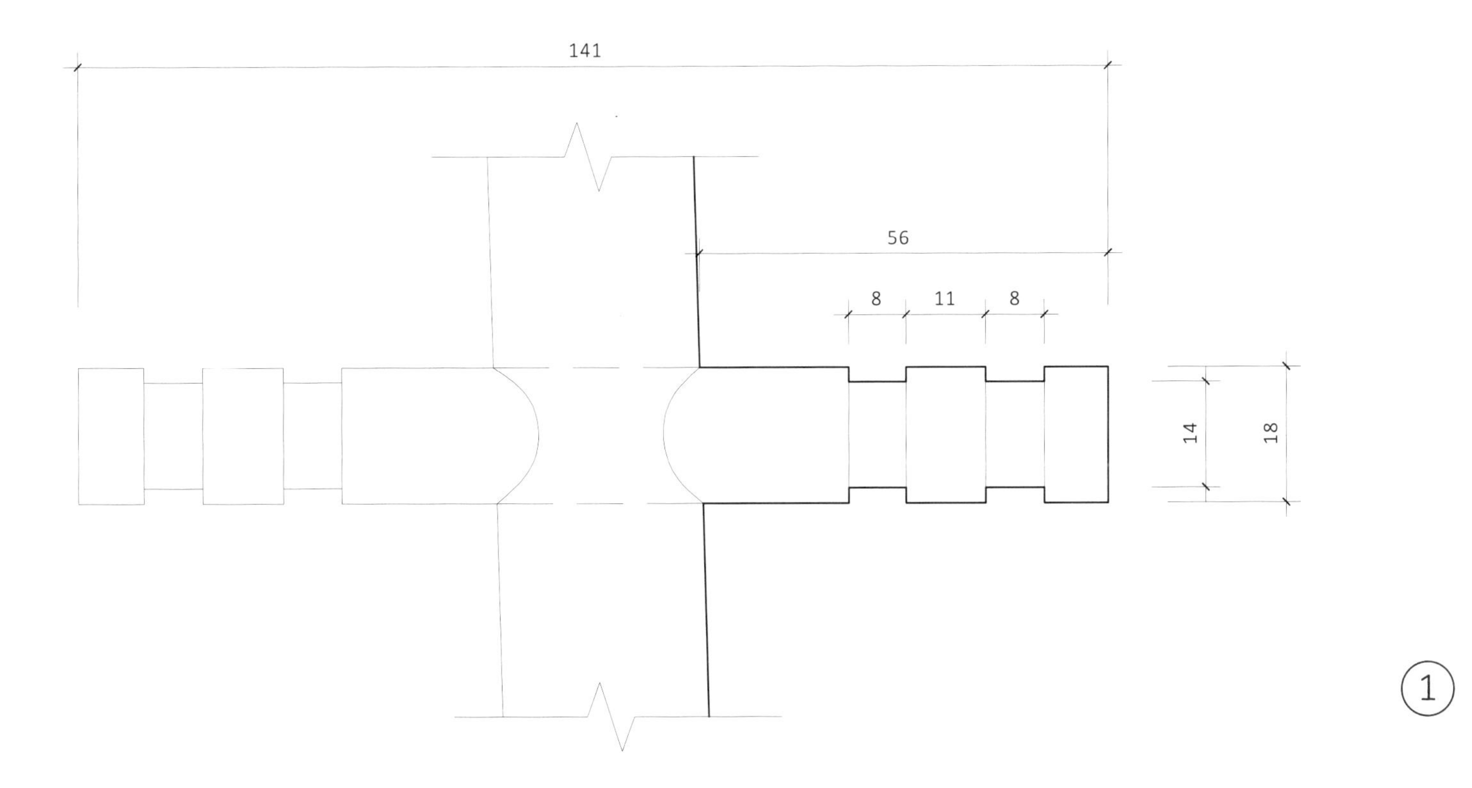

①

范雨佳 测绘
FAN YUJIA DELIN.

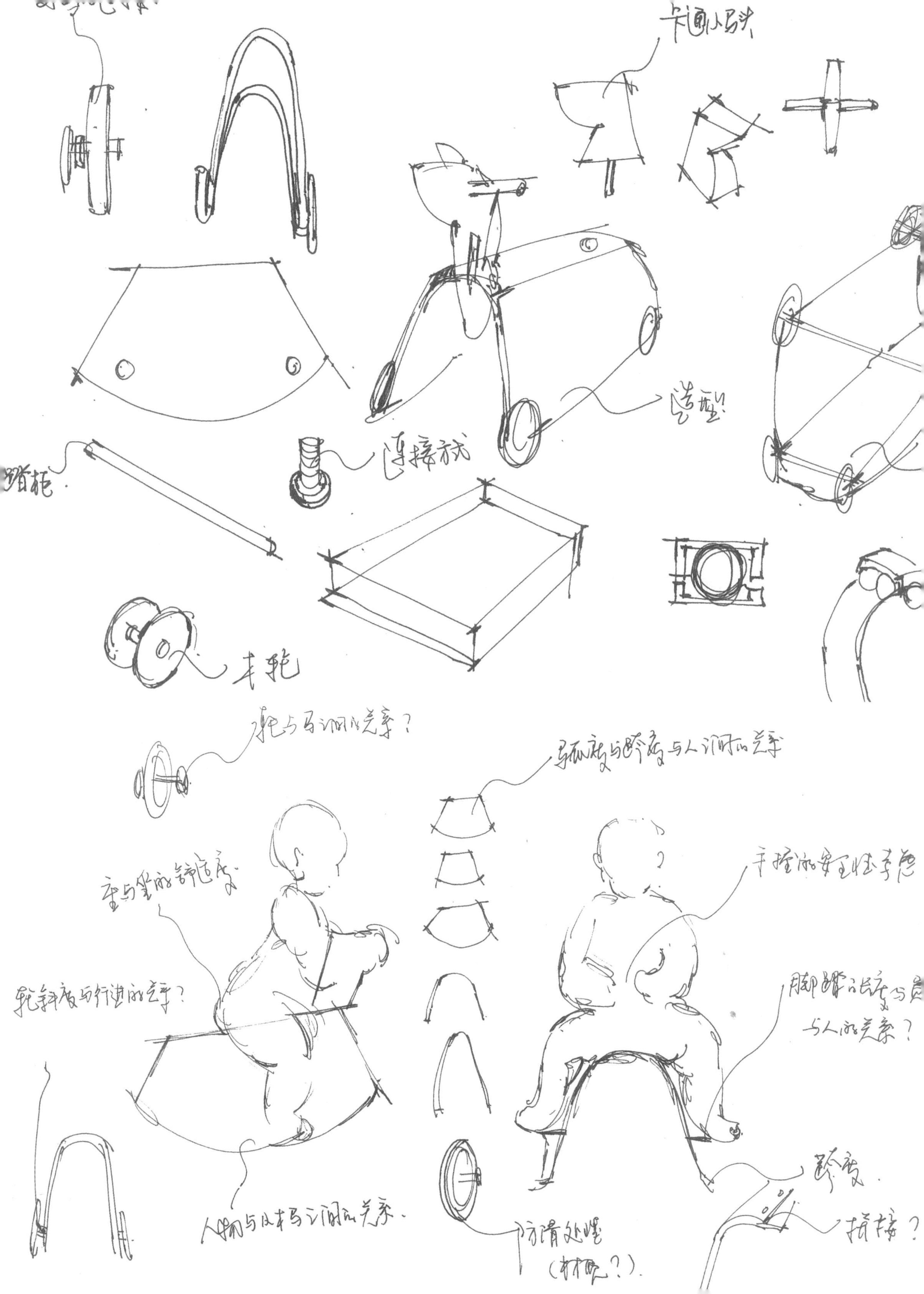

卡通小马头
造型!
连接杆
踏杆
木轮
轮与马之间的关系？
弧度与跨度与人之间的关系
手握的安全性考虑
座与坐的舒适度
轮斜度与行进的关系？
与人的关系？
人物与木马之间的关系
防滑处理
（材质？）
跨度
拼接？

小木马
The little horse

小木马儿童椅采用葡萄牙当地木材，既考虑儿童使用的趣味性，也关注使用的安全问题。可通过零件的拆装适应儿童不同年龄段的可持续使用需求。

规格 长 x 宽 x 高 404 x 725 x 498(mm)
设计师 郗鑫鑫
制作商 Camila Móveis Indústria Mobiliário Lda

Using Portugal local wood, the Little Horse takes into consideration child interest and safety in use. Sustainable use for children with different age can realize through dismantling or installing components.

Dimension length x width x height 725 x 440 x 418(mm)
Deigner Xi Xinxin
Maker Camila Móveis Indústria Mobiliário Lda

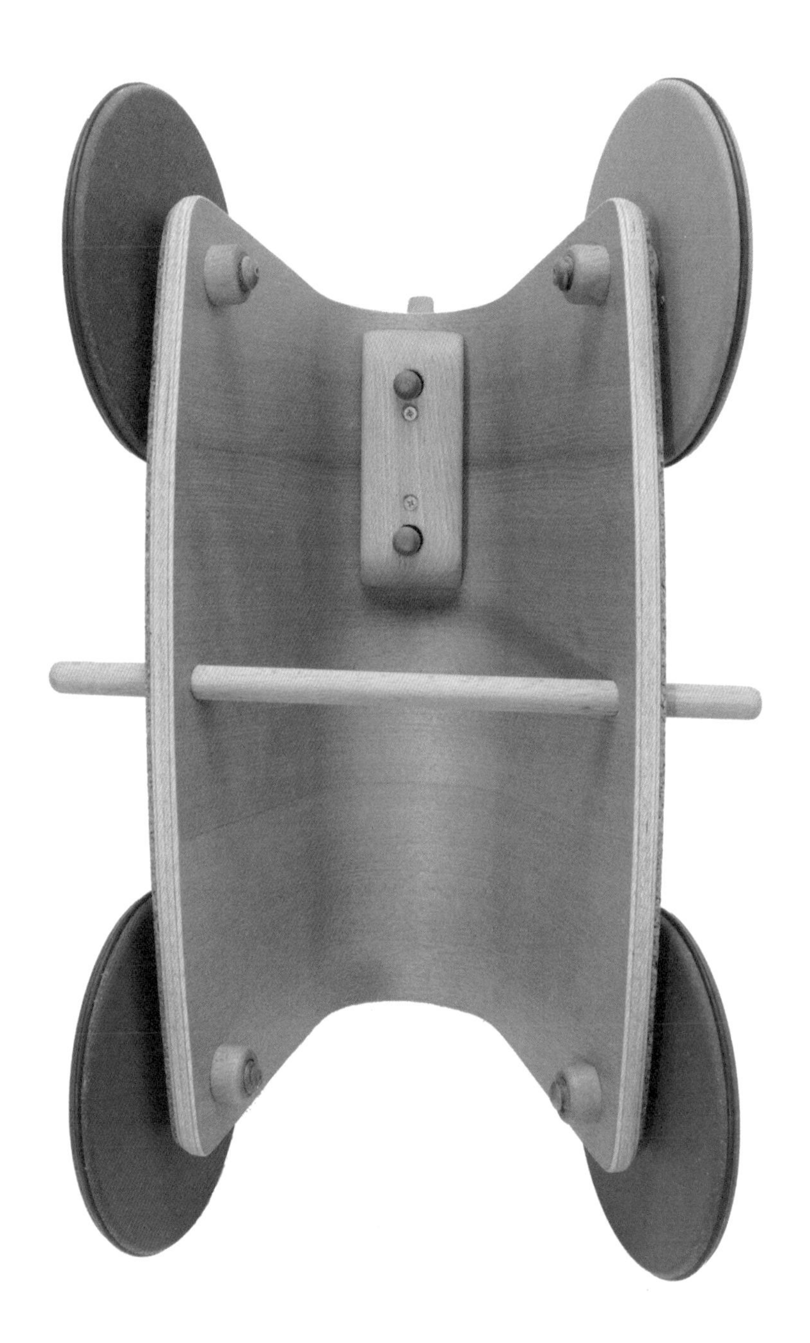

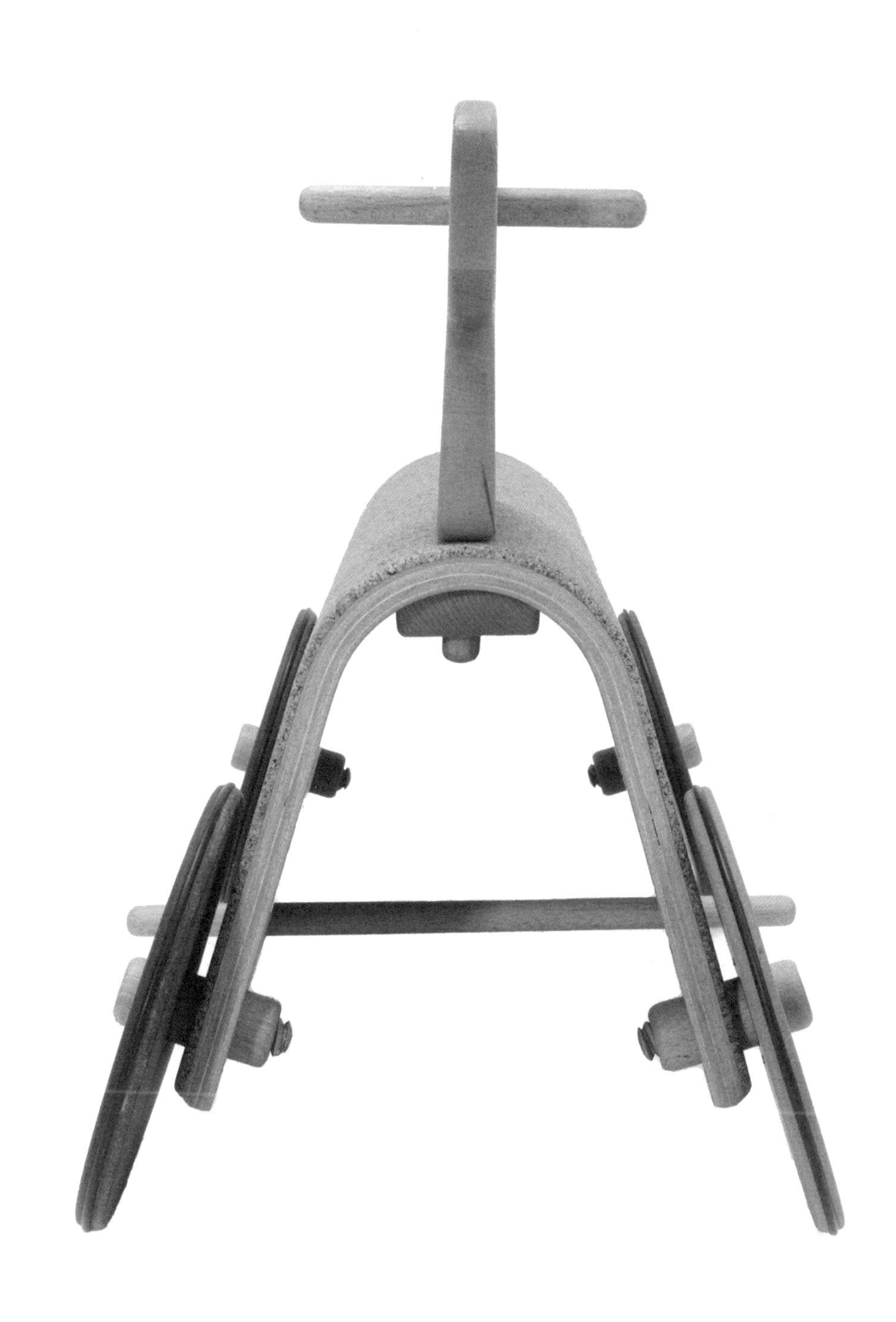

Camila
MOVEIS

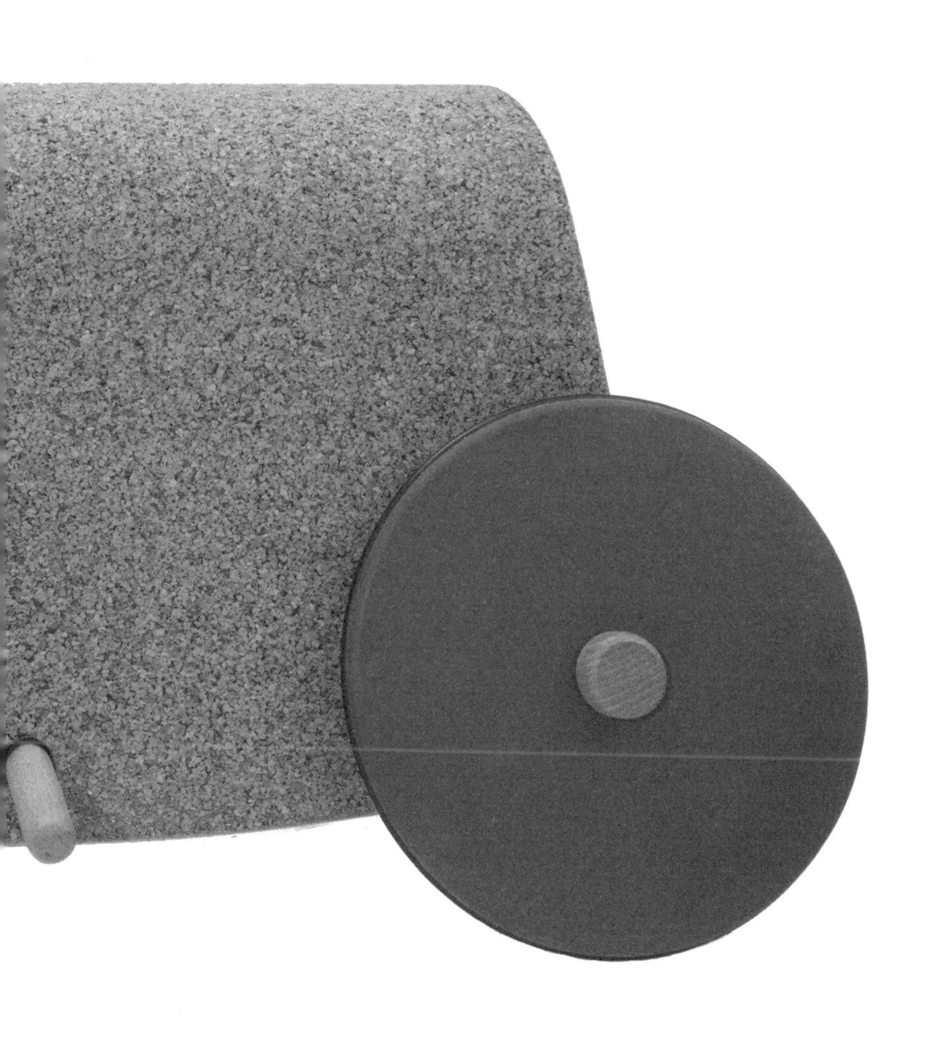

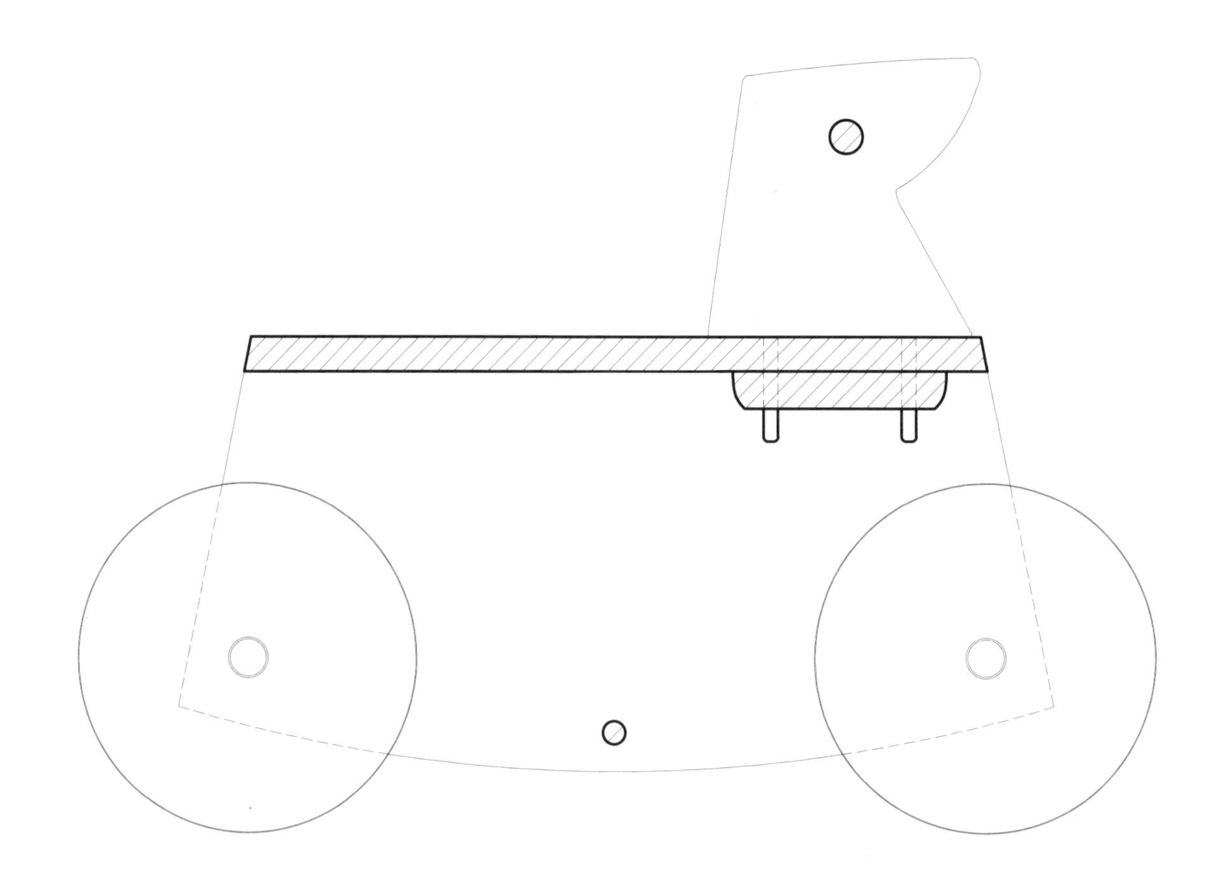

B剖面图 · SIDE AT B ·

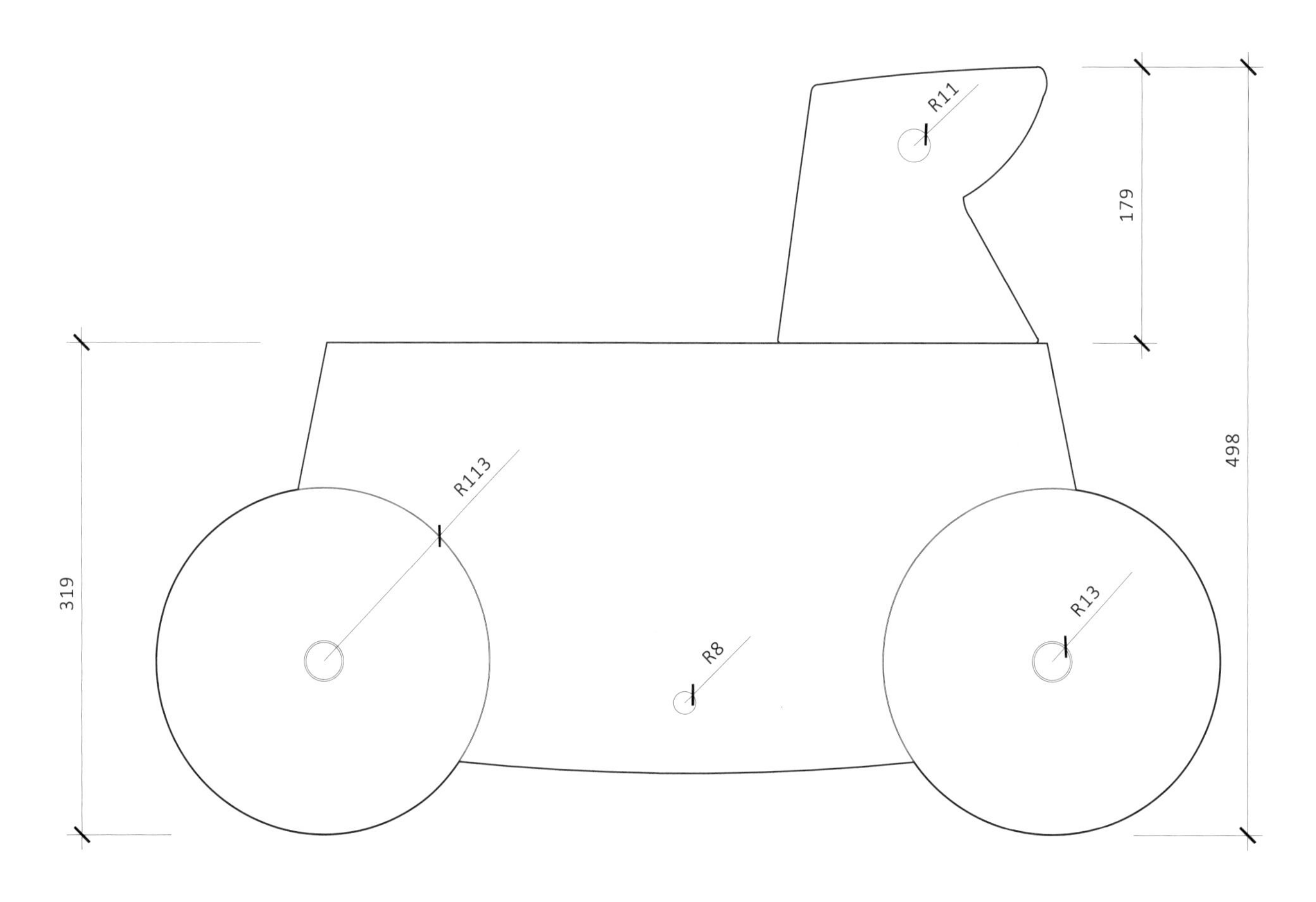

侧视图 · SIDE ELEVATION ·

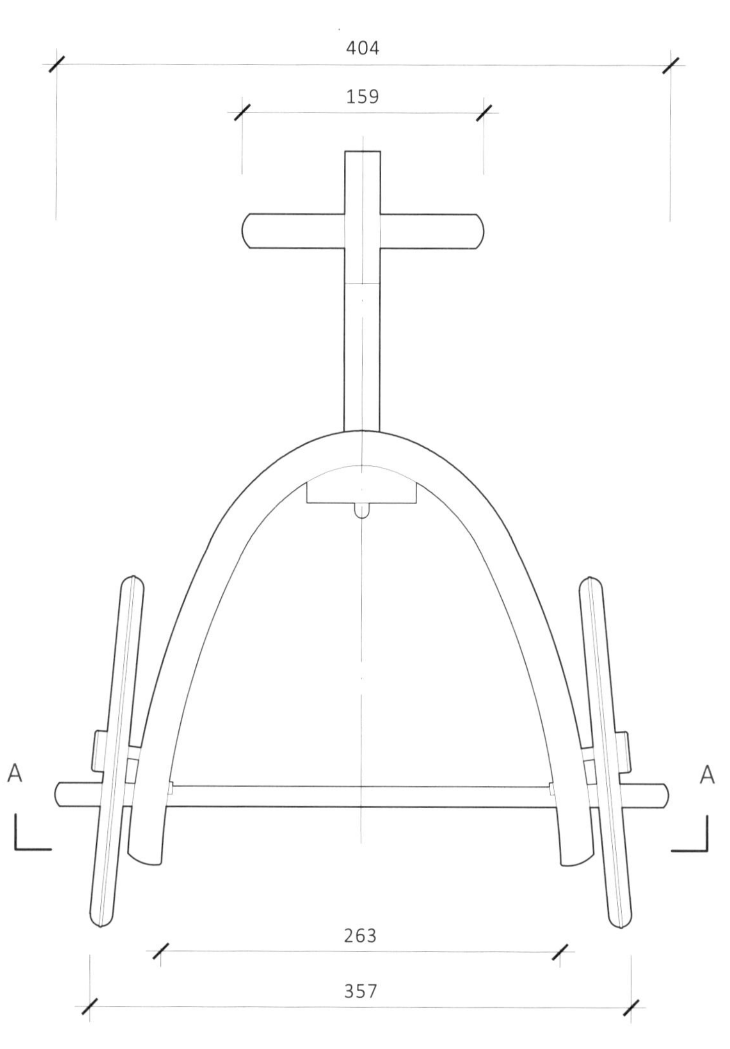

正视图 · FRONT ELEVATION ·

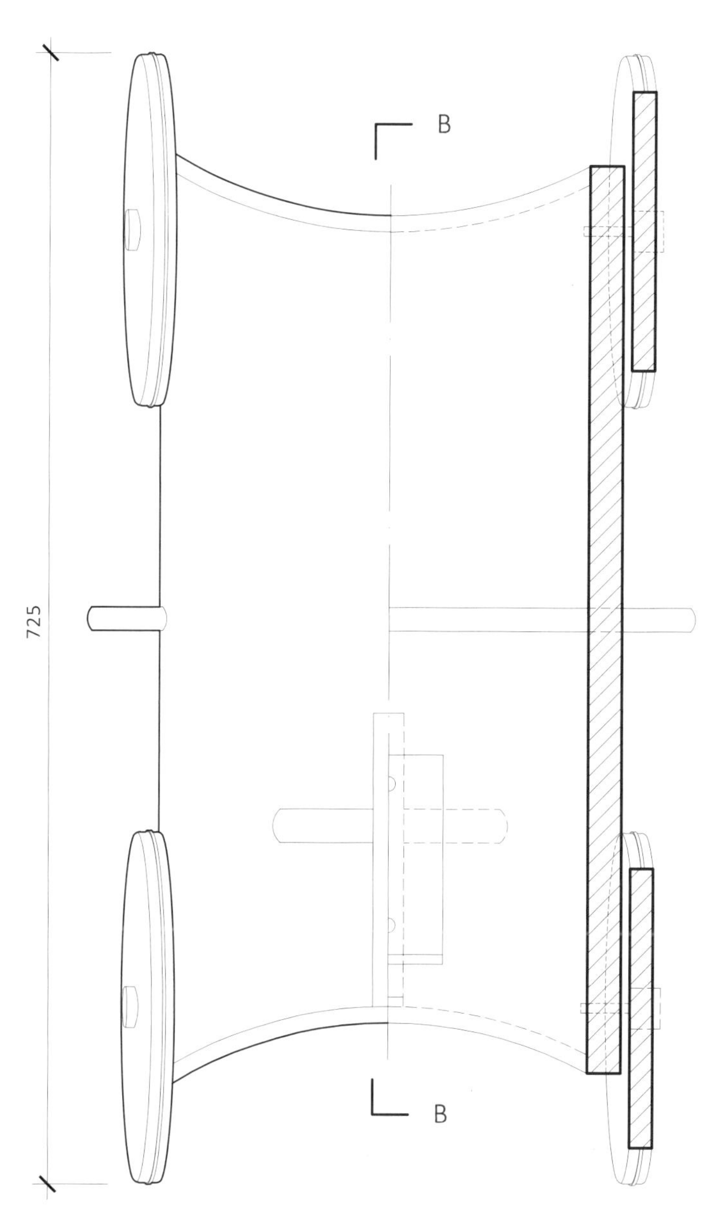

顶视图 · PLAN OF TOP ·

A刨面图 · PLAN AT A ·

赵帅 测绘
ZHAO SHUAI DELIN.

2016北京国际设计周
“中葡设计握手·家具篇——从北京到帕雷德斯”
The 2016 Beijing Design Week
"Handshake between Chinese and Portuguese Design. Furniture—
From Beijing to Paredes"

ART

艺

展示中葡设计交流 启迪文化传承创新

北京市人民政府外事办公室与北京建筑大学联合主办的2016北京国际设计周
“中葡设计握手 · 家具篇”（从北京到帕雷德斯）展览在北京歌华大厦开幕

陈静勇[1]

由北京市人民政府外事办公室和北京建筑大学联合主办，由歌华设计有限公司承办的2016北京国际设计周“中葡设计握手·家具篇”（从北京到帕雷德斯）展览开幕式于2016年9月9日在北京歌华大厦开幕。

展览布置从北京歌华大厦一层大堂开始，8个巨幅展览形象识别招贴悬挂在大堂上空，点燃了北京国际设计周热烈、浓厚的“设计节日”气氛。十三层DSC展厅中用中葡两国世界文化遗产做视觉引导，以“中国青年设计师驻场四季计划”工作过程为线索，通过实物、图片、文字、灯光等陈列设计手段，集中诠释出中葡两国设计的交流氛围。

展览开幕式于北京时间上午10点开始。莅临开幕式的主要领导和嘉宾有：北京市人民政府外事办公室副主任高志勇、葡萄牙驻华大使若热·托雷斯·佩雷拉、北京市国有文化资产监督管理办公室主任助理郑志宇、北京建筑大学校长张爱林、副校长汪苏、北京工业设计促进中心主任陈冬亮、文化部恭王府中华传统技艺研究与保护中心主任孙冬宁、北京歌华文化发展集团总经理李丹阳、北京市朝阳区实验小学书记王存友。参加开幕式还有2015年赴葡萄牙帕雷德斯市圆满完成中葡文化交流第二期“中国青年设计师驻场四季计划”（家具设计）专题项目的北京建筑大学师生设计师与北京市朝阳区实验小学教师设计师代表、中葡文化创意合作平台代表、北京建筑大学建筑与城市规划学院师生代表、中外媒体代表等。北京建筑大学校长张爱林为展览开幕做了热情洋溢的致辞。开幕式由北京建筑大学副校长汪苏主持。

2015年6~7月间以北京建筑大学为主的师生设计师团队一行10人，参加了在葡萄牙帕雷德斯市为期30天的第二期“中国青年设计师驻场四季计划”，主题为“家具设计”。师生设计师根据对葡萄牙当地文化的理解和感悟，借助帕雷德斯市5家家具企业，协同设计制作出面向中欧消费者的当代家具。

好的产品必须具备设计、工艺、市场化三要素。“驻场计划”项目使师生设计师们深度体验了“鼓励设计活动与工业生产同步、鼓励原创产品和内容创作、分享和传播，同时兼顾社会责任和可持续发展”的活动宗旨；感受到通过两国设计师与工匠的合作，探索创新设计与制造工艺和市场推广三位一体的新模式。

“驻场计划”活动受到中葡两国政府的高度重视。师生设计师在赴葡萄牙帕雷德斯市参加“驻场计划”项目时，受到葡萄牙时任总统卡瓦科·席尔瓦的亲切接见。师生设计师将北京建筑大学“家具构法与制作技艺研究所”（陈静勇教授工作室）与文化部恭王府中华传统技艺研究与保护中心联合主编出版的《中华传统技艺3. 2014小暑卷：明式家具传统制作技艺学术研讨会暨明氏十六品高仿作品展特辑》一书作为礼物送给了席尔瓦总统，席尔瓦总统与帕雷德斯市市长塞尔索·费雷拉分别签名留念。在葡期间，本校还与葡萄牙波尔图大学签署了合作办学协议，开启了与葡萄牙高校联合人才培养和国际交流的新阶段。

2016年，北京建筑大学作为文化部 教育部批准的全国首批57所“中国非物质文化遗产传承人

1 通信作者：北京建筑大学，教授，Email：chenjingyong@bucea.edu.cn

群研修研习培训项目”实施高校之一，与文化部恭王府管理中心共建了“中国非物质文化遗产研究院”；聘请校内外知名专家组成了“中国非物质文化遗产教学指导委员会”，在文化部非物质遗产司的指导下，开展“传统建筑营造技艺”“传统家具制作技艺”等传统技艺类别项目的保护与传承研究和人才培养。

借助知名的北京国际设计周平台，北京建筑大学在设计学学科将近年在“中国传统家具构法与制作技艺研究”中探索完成的部分明式家具“标准器”的复原作品与在本期“驻场计划”中完成的当代家具作品并置展出，以呈现中葡两国设计文化交流面貌，启发人们对文化传承与创新的思考。

展览开幕式结束后，与会人员共同参观了展览。领导、嘉宾在北京建筑大学关于传统家具技艺、传统木雕技艺学术出版物上赠言题字，与作品设计师生进行了交流和探讨。

北京建筑大学肇始于1907年的京师初等工业学堂，目前是北京市和住房城乡建设部共建高校、教育部“卓越工程师教育培养计划”试点高校，是一所具有鲜明建筑特色、以工为主的多科性大学，是“北京城市规划、建设、管理的人才培养基地和科技服务基地”“国家建筑遗产保护研究和人才培养基地”，是北京地区唯一一所建筑类高等学校。2016年5月，学校“未来城市设计高精尖创新中心”获批“北京高等学校高精尖创新中心”。

站在新的历史起点上，学校正按照“提质、转型、升级”的基本发展策略，围绕立德树人的根本任务，全面推进内涵建设，全面深化综合改革，全面实施依法治校，全面加强党的建设，持续增强学校的办学实力、核心竞争力和社会影响力，以首善标准推动学校各项事业上层次、上水平，向着建设国内一流、国际知名的有特色、高水平、创新型大学的宏伟目标奋进。

Show communities of design between China and Portugal Enlighten innovation of cultural inheritance

2016 Beijing international design week, CO sponsored by the Foreign Affairs Office of the people's Government of Beijing and the Beijing Architecture University

"The design of furniture, shake hands" (from Beijing to Paredes) the opening of the exhibition in Beijing Gehua building

Chen Jingyong[1]

On September 9th, 2016, the 2016 Beijing Design Week, the opening ceremony exhibition of "Handshake between Chinese and Portuguese Design. Furniture-from Beijing to Paredes" held in Beijing Gehua building. The exhibition jointly organized by Foreign Affairs Office of the People's Government of Beijing Municipality and Beijing University of Civil Engineering and Architecture undertook by Beijing Gehua Design Co., Ltd.

The exhibition layout started from the lobby of the first floor of Beijing Gehua Building. The eight huge identification posters of the exhibition image hung over the lobby, which lit up the enthusiastic and intense "design festival" atmosphere of Beijing Design Week. In the 13th story DSC exhibition hall, the world cultural heritage of China and Portugal used as visual guidance. With the clues of "Design-in-Residence Programmes of Chinese Young Designers in Four Seasons," the layout design methods including physical display, pictures, text, lighting, etc., centralized showed the design communication atmosphere between China and Portugal intensively.

The opening ceremony of the exhibition started at 10 a.m in China. The principal leaders and guests who presented at the opening ceremony were: Gao Zhiyong, deputy director of Foreign Affairs Office of the People's Government of Beijing Municipality, Jorge Torres Pereira, Portuguese Ambassador to China, Zheng Zhiyu, director assistant of State-owned Cultural Assets Supervision and Administration Office of the People's Government of Beijing Municipality, Zhang Ailing, President of Beijing University of Civil Engineering and Architecture, Wang Su, Vice-principal of Beijing University of Civil Engineering and Architecture, Chen Dongliang, director of Beijing Industrial Design Center, Sun Dongning, director of Chinese Traditional Skills Research and Protection Center of The Museum of Prince Kung's Palace, Li Danyang, general manager of Beijing Grhua Cultural Development Group, and Wang Cunyou, secretary of Beiging Chaoyang Experimental Primary School. People Participated in the Opening Ceremony were also the teachers and students of Beijing University of Civil Engineering and Architecture, who were the designers gone to Paredes, Portugal in 2015, and completed the second cultural exchange project, named "Design-in-Residence Programs of Chinese Young Designers in Four Seasons"(furniture design). Other groups included teacher designer representatives of Beijing Chaoyang Experimental Primary School, representatives of China- Portugal Cultural Creative Cooperation Platform, teacher and student representatives of School of Architecture and Urban Planning, Beijing University of Civil Engineering and Architecture, Chinese and Foreign Media Representatives, etc. Zhang Ailin, president of Beijing University of Civil Engineering and Architecture, gave a warm speech to the opening ceremony, which was hosted by Wang Su, the vice president of the university.

From June to July 2015, a team of 10 faculty and student designers led by Beijing University of Civil Engineering and Architecture took part in the 30-day second edition of the "Design-in-Residence Programmes of Chinese Young Designers in Four Seasons" in the city of Paredes, Portugal. The theme is "Furniture Design." The designers who depend on the comprehension and inspiration of Protuguse local culture made a cooperation with the five furniture company which in Paredes to design and make the modern furniture facing the consumer from China and Europe.

An excellent product must have three elements: design, technology, and marketization. The "Design-in-Residence Programmes" project enabled teachers and students to deeply experience the activities aim which is "Encouraging design activities in synchronizing with industrial production, encouraging to create, share and diffuse original product and content, and take into consideration social responsibility and sustainable development." Through the cooperation between designers and artisans from both countries, the teachers and students explored a new model of innovative design, manufacturing processes, and marketing.

The Chinese and Portuguese governments highly valued the "Design-in-Residence Programmes" activity. When the teacher and student designers participated in the project in Paredes, Portugal, they kindly received by the incumbent President of Portugal, Cavaco Silva. The designers gave the president Silva a book as a gift. The book was called the " Chinese traditional skills 3. volume slight heat 2014: traditional craft symposium of Ming style furniture and Ming shi sixteen high imitation works exhibition ". It jointly published by the Institute of Furniture Construction and Production of Beijing University of Civil Engineering and Architecture (Chen Jingyong Professor Studio), and Chinese Traditional Skills Research and Protection Center of The Museum of Prince Kung's Palace. President Silva and Colso Ferreira, the mayor of Paredes, respectively signed the book as a gift to teachers and students. During the Portuguese period, the university also signed a cooperation agreement with Portugal's Porto University and opened a new phase of joint personnel training and international exchange between the universities.

In 2016, Beijing University of Civil Engineering and Architecture is one of the first batches of 57 Implementation Universities in "Chinese Intangible Cultural Heritage Inheritance Training and Research Projects" approved by the Ministry of Culture and the Ministry of Education. And it jointly builds a "China Institute of Intangible Cultural Heritage " with the Museum of Prince Kung's Palace, which belonged to the Ministry of Culture. The Institute also hired well-known experts inside and outside of school. Under the guidance of the Intangible Heritage Protection Department, belonging to the Ministry of Culture, the intangible cultural skill inheritance and innovation research, as well as personnel training were carried out. These projects called "traditional architectural construction techniques" and "traditional furniture production skills."

With the help of the popular Beijing Design Week platform, Design Science discipline of Beijing University of Civil Engineering and Architecture exhibited some replicas of Ming Dynasty furniture "Standard device," explored by the "Research of traditional Chinese furniture construction techniques and production " in recent years, and contemporary furniture works completed in the this"Design-in-Residence Programs." The exhibition presented the cultural exchanges between China and Portugal and inspired people to think about cultural heritage and innovation.

After the opening ceremony of the exhibition, the participants visited the show. In the Beijing University of Civil Engineering and Architecture, the leaders and guests were pleased to give an inscription on the academic publications, whose theme was traditional furniture arts and traditional wood carving techniques. They also exchanged ideas and discuss with the student designers.

BUCEA Started from the Imperial Primary industry school that established in Qing dynasty in 1907. BUCEA is a multidisciplinary university which majors in engineering, and it has its features in architecture. Currently, BUCEA is a co-sponsored university with the People's Government of Beijing Municipality and Ministry of Housing and Urban-Rural Development and is a pilot university which is Excellent engineer training program by Ministry of Education. It is a talent training base and technology service base of Beijing urban planning, Construction, and Management, talent training base of research of dealing with changing the climate in Beijing, and talent training base of research on national architectural heritage protection. It is the only one architectural university of Beijing region. Beijing Advanced Innovation Center for Future Urban Design enter into University of Beijing Advanced Innovation Center. In May of this year, the school's "Beijing Advanced Innovation Center for Future Urban Design" approved as "Beijing Advanced Innovation Center."

The school, standing at a new historical starting point, is following the basic development strategy of "Quality Improvement, Transformation, and Upgrading." And the school focus on the fundamental tasks of strengthening moral education and cultivate people.It Comprehensively advances the connotation construction, deepens comprehensive reforms, implements the managing schools by law, and strengthens party building. And it continues to enhance the school strength, core competitiveness, and social influence to promote all the school's careers with the highest standards. And move towards the grand goal of "building the school into a domestic first-class, internationally-renowned, distinctive, high-level, and innovative university by the time the 100th anniversary of the establishment of the university in 2036".

2016年9月9日，葡萄牙驻华大使若热・托雷斯・佩雷拉莅临展览开幕式并签名留言
On September 9, 2016, Portuguese Ambassador to China Jorge Torres Pereira visited the opening ceremony and signed .

1 Communication author, Beijing University of Civil Engineering and Architecture, Professor, Email:chenjingyong@bucea.edu.cn

北京市人民政府外事办公室副主任高志勇、北京建筑大学校长张爱林与葡萄牙驻华大使若热·托雷斯·佩雷拉在展览现场亲切交流。
Gao Zhiyong, deputy director of the Foreign Affairs Office of the People's Government of Beijing Municipality, and Zhang Ailin, president of Beijing University of Civil Engineering and Architecture, exchanged ideas with the Portuguese ambassador to China, Jorge Torres Pereira.

北京建筑大学副校长汪苏主持展览开幕式并与主创人员交流
Wang Su, vice president of Beijing University of Architecture, host the opening ceremony of the exhibition and communicate with creators.

北京建筑大学党委副书记张启鸿莅临展览与主创人员进行交流
Zhang Qihong, deputy party secretary of Beijing University of Civil Engineering and Architecture, came to the exhibition and communicate with creators.

文化部恭王府中华传统技艺研究与保护中心主任孙冬宁莅临展览并与主创人员交流
Sun Dongning, director of Chinese Traditional Skills Research and Protection Center of The Museum of Prince Kung's Palace, came to the exhibition and communicate with creators.

北京建筑大学“传统家具构法与制作技艺研究所”主持人陈静勇教授为葡萄牙驻华大使若热·托雷斯·佩雷拉和校长张爱林汇报明式家具“标准器”复原研究情况。
Professor Chen Jingyong, the host of the "Institute of Traditional Furniture Structure Method & Craftsmanship" of Beijing University of Architecture, reported on the replication of the "standard device" of Ming furniture for the Portuguese ambassador to China, Jorge Torres Pereira, and the headmaster Zhang Ailin.

第二期“中国青年设计师驻场计划”设计师团队领队北京建筑大学副教授杨琳在展览现场接受媒体采访。
"Phase || of Design-in-Residence Programs" designer team leader Yang Lin, Associate Professor of Beijing University of Civil Engineering and Architecture, interviewed by the media at the exhibition site.

中葡设计握手 家具篇
2015

家具文化的传承与创新探讨

2016北京国际设计周："中葡设计握手·家具篇"（从北京到帕雷德斯）
策展设计

林则全[1] 李明军[2] 闫卓远[3] 范雨佳[1] 杨琳[1] 陈静勇[T]

摘要

以"中葡设计握手·家具篇"展览为例，探讨家具文化的传承与创新。通过对比中葡文化符号和梳理中葡家具交流历程，引出此次策展设计立足中华民族的优秀传统文化内在精神，开展国际合作与交流协同创新的设计理念；新时代坚定走传统文化的传承与创新设计之路；将传统文化融入孩童教育，树立文化自信是传承与创新的重中之重。

关键词

设计学；家具文化；中葡"驻场计划"；策展设计；传承与创新

1 北京建筑大学设计艺术研究院　2北京歌华设计有限公司　3中国科学技术馆
T 通信作者：北京建筑大学
"传统家具构法与制作技艺研究所"，教授，Email:chenjingyong@bucea.edu.cn

A Discussion on the inheritance and innovation of furniture culture

2016 Beijing International Design Week: "Handshake between Chinese and Portuguess ·Furniture, From Beijing to Paredes" exhibition design

Lin Zequan[1] Li Mingjun[2] Yan Zhuoyuan[3] Fan Yujia[1] Yan Lin[1] Chen Jingyong[T]

ABSTRACT

The inheritance and innovation of furniture culture discussed by the example of "Handshake between Portuguese and Chinese Design. Furniture." Comparing cultural symbols and communications between Chinese and Portuguese, we conclude that this curatorial design based on the inner spirit of the excellent Chinese traditional culture.It develops international cooperation and collaborative and innovative design concept. The New Era Stands Firmly on the Inheritance and Innovative Design of Traditional Culture. Building the confidence of traditional culture and putting it into the education of children is significant.

KEYWORDS

Design Science, Furniture culture,
"Design-in-Orsidence Programs" of Chinese and Portuguese,
Exhibition design, Inheritance and Innovation

1 Academic of design and art of Beijing University of Civil Engineering and Architecture
2 Beijing Gehua Design Co.,Ltd
3 China Science and Technology Museum
T Communication author, Beijing University of Civil Engineering and Architecture,
Institute of Traditional Furniture Structure Method & Craftsmanship, Professor, Email:chenjingyong@bucea.edu.cn

北京故宫
The Forbidden City

文化符号

在5000年前亚洲大陆的中原地带诞生了一个国家——中国。她所孕育的人民饱经沧桑，虽然经历了多次朝代更迭，但在民族交融中始终保持着中华文化的核心基础。五千年的发展，将中华文化的博大精深，绚烂多彩展现给了世界。中华文化在语言文字、文学礼制、音乐绘画等方面都有着深厚的底蕴，发展出了诸如京剧、书法、国画、工艺美术、中医、武术等国粹。随着时间的推移，中国的传统文化越显得稳重厚实，文化的繁荣发展使得中国闪耀在世界的东方，也让中国走进了世界舞台的中央。

位于伊比利亚半岛的葡萄牙是欧洲最为古老的国家之一。友好热情的葡萄牙人在大航海时代为地理大发现做出了长足的贡献。葡萄牙也以其丰富的文化遗产和建筑闻名世界。作为一个狂热的足球国度，独具拉丁风味的葡萄牙足球风靡全球。在音乐、饮食等方面葡萄牙也诞生出了诸如法多的音乐形式和美味多样的鳕鱼做法等。葡萄牙像是一片最亮眼的贝壳镶嵌在欧洲大陆的最西端，散发着她的无穷魅力。

葡萄牙的国土面积不大，目前却有着15处世界遗产[1]，有像贝伦塔那样的防御建筑，见证葡萄牙辉煌的历史；也有如阿尔科巴萨修道院等的宗教建筑，是中世纪哥特建筑的代表作品；更有像波尔图那样的老城市历史中心，他的历史甚至可以追溯到葡萄牙建国之前。

中国目前有着52处世界遗产[2]，长城作为中国古代巨大的军事防御工程的代表被写进世界遗产名录。长城的建造贯穿中国历史，同样也成为中国历史的见证者。北京故宫集中国古代宫廷建筑的精华，展现了中国辉煌深厚的文化。天坛是中国保存下来的最大的祭祀建筑群，代表着中国人“天人合一”的思想观念。

两国的世界遗产不仅代表着忆往日的璀璨，更代表着向未来的寄望。芳华将至，未来更需要年轻的一代人加入保护世界遗产的队伍。

1 葡萄牙世界遗产名录
2 中国世界遗产名录

里斯本
Lisbon

Cultural Symbols

A country which named China bored in the central plains zone of Asia in 5000 years ago. Although the people in China experienced many times of dynasty change, they still keep the base of Chinese culture with ethnic integration. The extensive and profound of Chinese cultures shown to the world with the development of the period of 5000 years. There is a broad strength in the quintessence of Chinese cultures, such as language, etiquette, music, and painting. Moreover, that developed many kinds of traditional cultural, such as Beijing opera, Chinese calligraphy, Chinese painting, arts and crafts, Chinese medicine and martial art. With time following, increased dignity can be seen in Chinese traditional culture. China has shone in the east of the world and joined the heart of the world with a rich culture.

Portugal is one of the ancient countries of Europe which located in the Iberian Peninsula. Portuguese made a significant contribution to the Big Maritime Navigation Era. Also, Portugal is well known for cultural heritage and architecture. As a fanatic football country, Portuguese football with Latin style is famous around the world. Unique music and food were bored in Portugal. There are also many creations in music, diet. For instance, Portuguese folk song named Fado and a variety of methods of cooking codfish have invented. Portugal is like the most dazzling shell inlaid at the western end of continental Europe. It is showing her infinite charm to the world.

Portugal has a small land area and currently has 15 world heritage sites[1]. For example, there is a defense construction named Belem Tower which witnesses the glorious history of Portugal, a religious architecture called Monastery of Alcobaca which is the symbol of Gothic architecture in the Middle Ages and an old city named Porto which even establish before old Portugal.

At present, China has 52 World Heritages[2]. The Great Wall wrote on the World Heritage list as a substantial Military defense engineering in ancient China. It becomes a witness to Chinese history. The essence of palace architecture in Ancient Chinese is The Forbidden City, and it displays the beautiful culture of China. Temple of Heaven which symbolizes the traditional Chinese thinking of "harmony between man and nature" is the most significant group of sacrificial buildings.

The world heritages of two countries represent not only the memories of past days' resplendent but also the hope of the future. A good time is approaching, and there needs the younger generation to protect the World Heritage in the brilliant future.

1 Portugal World Heritage List http://whc.unesco.org/en/statesparties/pt
2 China World Heritage Listhttp://whc.unesco.org/en/statesparties/cn

中葡家具文化交流

大约在17世纪到18世纪，欧洲的海上贸易为欧洲带回了许多东方的瓷器、丝绸和家具。这个契机领导了盛行欧洲的“中国风”。早在法国路易十五时期，中国的大漆装饰在欧洲广泛流行。受到中国华丽的东方情调影响，大胆采用自由曲线的洛可可风格家具解构了中国园林中的建筑，将中国园林中的塔、窗格等元素用在了家具上。在当时的欧洲社会，仰慕和追求中国的社会生活和华丽风格的漆器家具成为新风尚，这也是中国家具对欧洲家具最早的影响。

随着欧洲工业化进程发展，19世纪末，原先那种华丽的风格不再是社会追求的时髦。技术的变革，也带来了审美的变革，从富丽堂皇的装饰风格变为实用主义的机器美学。简练的明式家具随即在欧洲的市场中脱颖而出，在其后的百年间对各个欧洲国家的家具设计产生了或多或少的影响，直到现在也能在从这些家具中发现中国明式家具的影子。

欧洲的海上贸易不仅给他们带去了“中国风”，也让一些欧洲传教士留在了中国，使得一些欧洲的家具传入了中国。中国的家具在“康乾盛世”的时代不仅继承了传统，还主动吸收欧洲风格，产生出中西合璧的设计风格。在一些清式家具的装饰纹样中出现了洛可可风格和巴洛克风格中典型的扇形贝壳纹、涡卷纹等。这也是中国家具由简洁的明式家具向华丽的清式家具转变的一个重要的影响因素。

随着中西方交流更加的频繁，西为中用的社会背景下，民国时期家具融汇多种西方古典家具风格，表现出中西合璧风格的杂糅统一。不只是西方的古典家具，其后流行的西方现代主义风格家具也随着时代大潮流，传入中国，这些家具受到了当时时尚界人士的青睐。

最早输入西方的瓷器，是由葡萄牙人所领导的。青花瓷作为一种大规模商品的传入和消费，首先成为葡萄牙的社会风尚，继而形成欧洲的社会风尚。由此开始，其后的欧洲“中国风”在葡萄牙里斯本桑托斯宫体现得淋漓尽致，在这座宫殿中的一个房间中的金字塔式圆拱的三角形四边由260件青花瓷盘覆盖。

在其后的日子里青花瓷一直影响着葡萄牙。由于葡萄牙是直接与中国交易青花瓷的国家，青花瓷不仅是宫廷享用的奢侈品，更是逐渐成为寻常百姓家的日用品。走在葡萄牙的街头，随处可以看见由青花瓷装点的建筑外立面和街头雕塑。在现代的一些家具设计和产品设计中也会运用青花瓷的设计元素来表现。

Cultural exchange of furniture between China and Portugal

From seventeenth Century to Eighteenth Century, European maritime trade brought back much Eastern porcelain, silk, and furniture for Europe. This opportunity led to the "Chinese wind" that prevailed in Europe. As early as the French Louis fifteen period, Chinese lacquer decoration is widely popular in Europe. Influenced by the beautiful Oriental sentiment of China, the Rococo style furniture with free curve boldly deconstructed the architectures in Chinese gardens, and the elements such as pagodas and panes in Chinese gardens used on furniture. In the European society at that time, admiration and pursuit of Chinese social life and gorgeous style lacquer furniture became a new trend. It is also the earliest influence of Chinese furniture on European furniture.

With the development of European industrialization, at the end of the Nineteenth Century, the original style was no longer the fashion of social pursuit. The transformation of technology had also brought aesthetic changes. It was from the magnificent decoration style to the pragmatism of the machine aesthetics. Concise Ming-style furniture stood out in the European market. In the following hundred years, it has had more or less influence on the furniture design of various European countries. Until now, the imprints of Chinese Ming-style furniture can also found in this furniture.

Europe's maritime trade not only brought them the "Chinese style," but also left some European missionaries in China, so that some European furniture introduced into China. Chinese furniture in the "flourishing age" era not only inherited the traditional European style but also took the initiative to absorb the design style of the Chinese and Western. In some of the decorative patterns of the Qing style furniture, the Rococo style and the typical scallop and scroll patterns in the baroque style had appeared. It is also an important influence factor for the transformation of Chinese furniture from simple, clear style furniture to luxuriant style furniture.

With more frequent exchanges between China and the west, and at the social background that west things used served China, the period of the Republic of China furniture integrated various western classical furniture style. It was showing a unified form of Chinese and Western hybridity. Not only the traditional furniture of the west but also the modernist furniture of western was introduced into China with the trend of the times. Some fashionable people favored this furniture at that time.

The Portuguese led the earliest porcelain which imported into the West. As a kind of large-scale commodity, blue and white porcelain was introduced and consumed, and first became social customs of Portuguese, then formed in Europe. From then on, the subsequent "China Wind" in Europe reflected in the Santos House in Lisbon, Portugal. The triangular sides of the pyramidal arch in a room in this palace covered with 260 blue and white porcelain plates.

Afterward, the blue and white porcelain had been affecting Portugal. Because Portugal was a country that deals with blue and white porcelain directly with China, blue and white porcelain was not only a luxury that palace enjoys, but also a daily necessity of common people's home. Walking on the streets of Portugal, the architectural facades and street sculptures decorated with blue and white porcelain can see everywhere. In some modern furniture design and product design, the design elements of blue and white porcelain will also use.

《绅士与橱柜制作者指南》 奇彭代尔 1754
"The Gentleman and Cabinet Maker's Director",Thomas Chippendale,1754

《清代家具》 田家青 2012
"Furniture in Qing Dynasty",Tian Jiaqing,2012

参考文献

[1]程庸.中国艺术品影响欧洲三百年之"中国风"与中国家具[J].家具,2010(06):52-57.
[2]王慧敏. 从"紫檀有束腰西洋装饰扶手椅"看清代中期家具的中西融合形态[D].南京艺术学院,2007.
[3]万明.明代青花瓷西传的历程:以澳门贸易为中心[J].海交史研究,2010(02):42-55
[4]Pássaro, Cátia. (2015). *A Chinoiserie no mobiliário português do século XVIII: história, arte e mercado.* Dissertação de Mestrado em Gestão de Mercados da Arte. Lisboa: Departamento de Sociologia e Políticas Públicas do Instituto Universitário de Lisboa.

“中葡设计握手·家具篇”策展设计

Design of "Handshake Between Chinese and Portuguese Design. Furniture " Exhibition

列入2016北京国际设计周的“中葡设计握手·家具篇（从北京到帕雷德斯）”主题展览与中外交流会活动是以促进国际文化交流为目的，以中国家具文化为内核，以家具设计为载体，以2015中葡文化交流第二期“中国青年设计师驻场四季计划”（家具设计）专题项目交流成果为展现，构成这次展览与交流活动。此次展览是由北京市人民政府外事办公室和北京建筑大学主办，由北京歌华设计有限公司承办的带有国际交流性质的展览。展览主要展示北京建筑大学近年研究复制的8件明式家具，分别为“圈椅”“灯挂椅”“六方形南官帽椅”“玫瑰椅”“四出头官帽椅”“夹头榫带屉板平头案”“翘头案”“一腿三牙罗锅枨方桌”和中葡“驻场计划”中由中国青年设计师2015年在葡期间设计的10件家具，分别为“总统椅”“总统桌”“缠绕边桌”“咖啡几”“妆·纳梳妆台”“小木马童车”“分·格空间架”“一方凳”“清风鞋柜”“融错·空间衣架”。

展陈策划上，总体布局取把握中华民族的优秀传统文化内在精神，开展国际合作与交流协同创新永续传承之意，展开设计。

此次展览位于北京歌华大厦A区13层DSC展厅。进入展厅需经过歌华大厦1层大堂，再乘电梯抵达13层展厅。展厅平面呈现为东西长向矩形平面空间，西、北两向幕墙外窗，中间两根立柱。展览设计将展厅分为东部的礼仪空间和西部的主展览空间。

礼仪空间的设计呈半围合状。位南朝北的是致辞剪彩区，以避开正对入口咖啡厅，使礼仪空间的形成不被周边因素所影响；北侧为2015年中葡“驻场计划”交流过程展示区，包括两国代表性的世界文化遗产、设计师团队在葡期间的行程照片、葡萄牙时任总统阿尼巴尔·卡瓦科·席瓦尔和帕雷德斯市市长塞尔索·费雷拉分别为北京建筑大学联合文化部恭王府中华传统技艺研究与保护中心编著出版的《中华传统技艺3·2014小暑卷：明式家具传统制作技艺学术研讨会暨明氏十六品高仿作品展特辑》签字留念；东侧为中国传统家具代表性榫卯结构的展示区，一方面是与东临的办公区域做分隔，另一方面是展示中国传统家具制作技艺的精华代表。

主展览空间呈环绕状展示。外圈为2015中葡文化交流“驻场计划”活动设计师团队在葡期间设计的10件家具，内圈为北京建筑大学“传统家具构法与制作技艺研究所”研究复制的8件明

The exhibition named "Handshake Between Chinese and Portuguese. Furniture" was purposed to promote the exchange of international culture in 2016 Beijing design week. The core was Chinese furniture culture, and it showed as furniture design with Phase II of Design-in-Resident Programs. This exhibition hosted by Foreign Affairs Office of the People's Government of Beijing Municipality and Beijing University of Civil Engineering and Architecture and undertook by Beijing Gehua Design Co., Ltd. The main displays were the eight pieces of Ming-Style furniture of rehabilitation design by BUCEA in recent years. Those were Round-backed armchair, Lamp-hanger chair, Southern official's hat armchair, Rose chair, Armchair with four protruding ends, Elongated bridle joint flat-top recessed-leg table with drawers, Head chair and Three spandrels humpbacked stretcher Square table" respectively. Moreover, the ten pieces of furniture were designed by ten young Chinese designers during the Programs in Portugal in the year of 2015. Those were President chair, President table, Binding, Coffee table, Sensation, Hobbyhorse, Separate space, A cube, Breeze and The interlaced dimensions respectively.

Concerning exhibition planning, the meaning of entire layout was the inner spirit of Chinese traditional culture and eternal inherited with international cooperation.

This exhibition located in 13th floor DSC exhibition hall, Beijing Gehua Tower Block A. There was a lobby in the first floor, one should through over it before arrive the DSC exhibition hall.Exhibition layout has a rectangle space from east to west. Besides, the walls in the north and west are both glass curtain wall. Also, there were two pillars in the center of the hall. Exhibition design divided exhibition hall into two parts. The ceremonial space was in the east, and the central exhibition space was in the west.

The design of the ceremonial space was semi-enclosed. To avoid the annoy by the entr-ance cafe, it set the ribbon cutting area in the South and face to the north. In the north part, there were processes of programs, including the picture about some typical world heritages and traveling photos that the design team in Portugal. What is more, there placed books which compiled by Chinese traditional skills research and protection center in the museum of Prince Kung's palace and BUCEA. It named Chinese traditional skills 3 · 2014 volume slight heat: traditional craft symposium of Ming style furniture and Ming Shi sixteen high Imitation works exhibition that signed by the president of Portugal AníbalCavaco Silva and the mayor of Paredes Colso Ferreira. The tenon structure display area set in the eastern of the exhibition were separated with the office

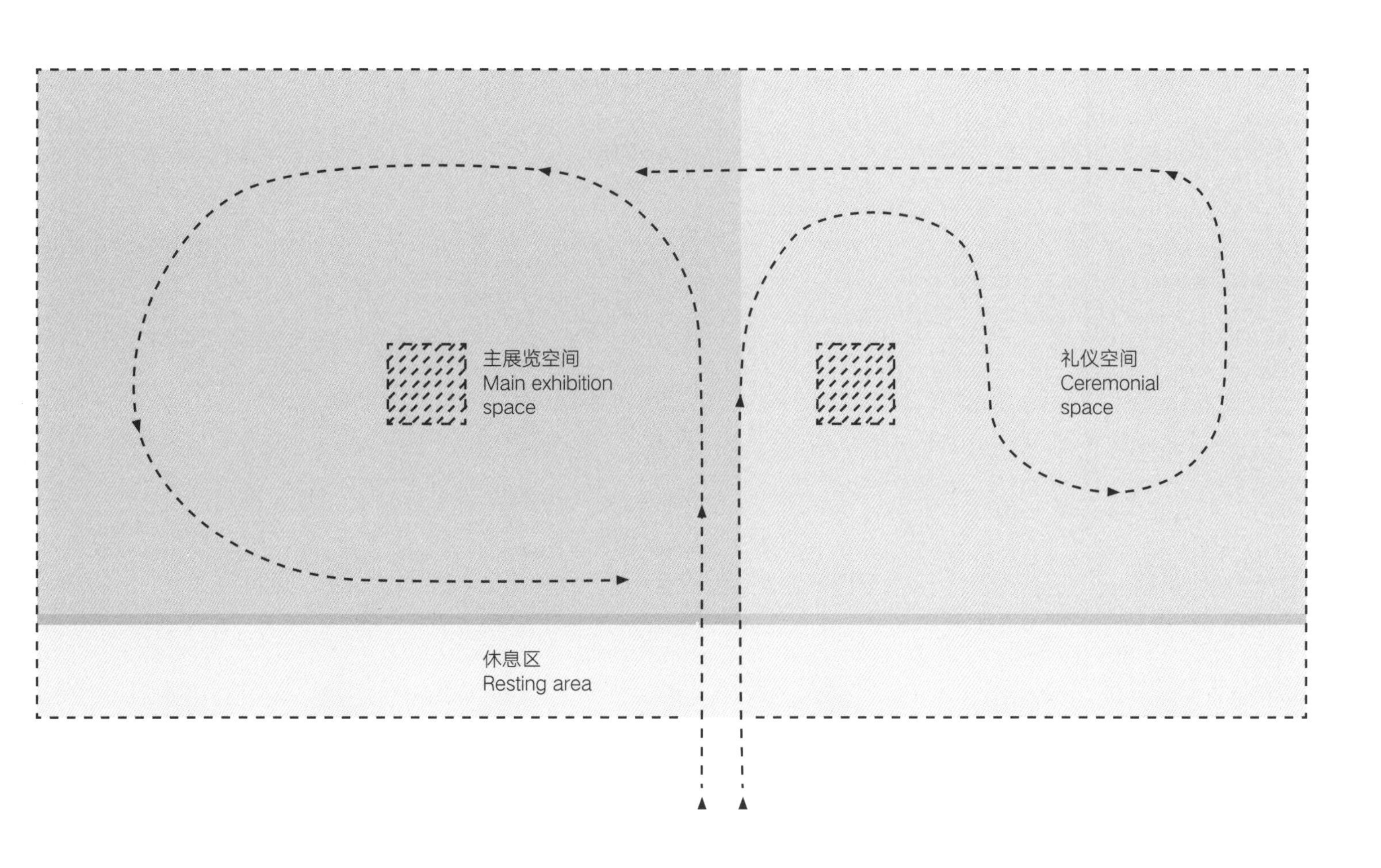

展览流线及空间布局
Exhibition streamline and space layout

式家具。

展厅西、北两面幕墙外窗遮光百叶窗关闭，与外界环境隔离，营造暗环境。同时，地面整体铺设浅灰色地毯，使得参观者将目光集中在展品上。

家具置于白色展台上。介绍展布位于展品后方，从屋顶钢架上悬吊下来，其状“贯通天地”，有效利用空间通高，以期延长视觉，充分表达。布展内容从上到下依次展示“家具照片”“家具品名”“作者照片和介绍”“设计理念”“设计场景照片”“设计草图”。展现家具设计过程，引入设计师的创新情境。

主展览空间的灯光设计整体偏暖暗，中强聚光直射家具，弱光漫洒在旁侧及展布上，使得视觉集中于家具本身，用以烘托以展品为核心的整体氛围。而礼仪空间，其灯光设计较亮，以便于人流聚集和疏散。

展厅入口区设签到台，剪彩区域矗立中葡两国国旗，铺红地毯。依照剪彩日活动要求展场灯光设置和展览流线都做了单独设计。

and showed the symbol of traditional Chinese furniture making skills.

The central exhibition space was in a hoop-like display. Outside displays were the ten pieces of furniture which were designed by young Chinese designer, and the other eight pieces of furniture were inside which were researched by Traditional Furniture Architecture and Production Technique Institute in BUCEA. Closing all shutter to isolated the exhibition from natural light, at the same time, the floor covered with a light grey carpet that the visitors could focus on the exhibits.

Closing all shutter to isolated the exhibition without natural light, at the same time, and the floor were carpted in a light grey that the visitors could focuse on the exhibits.

Furniture set on the white box. Introductions of furniture were behind the exhibits. From the beam frame to the ground, it was efficiency in space using. From top to bottom, there were Furniture picture, Furniture name, Photo, and the introduction of designer, Design idea, working picture and Sketch which brought the visitors into the sense of designer imagination.

Warm, slight spotlight touch the furniture directly, and the glimmer surrounded so that audiences could concentrate on the furniture themselves. Compare with the central exhibition space, and the light was intense which is convenient for congestion and evacuation of people in ceremonial space.

There was a reception table in the entrance. The ribbon cutting area carpeted in red, and there are both China and Portugal national flags. According to the requirements of the opening day, lights set and visit line had individually designed.

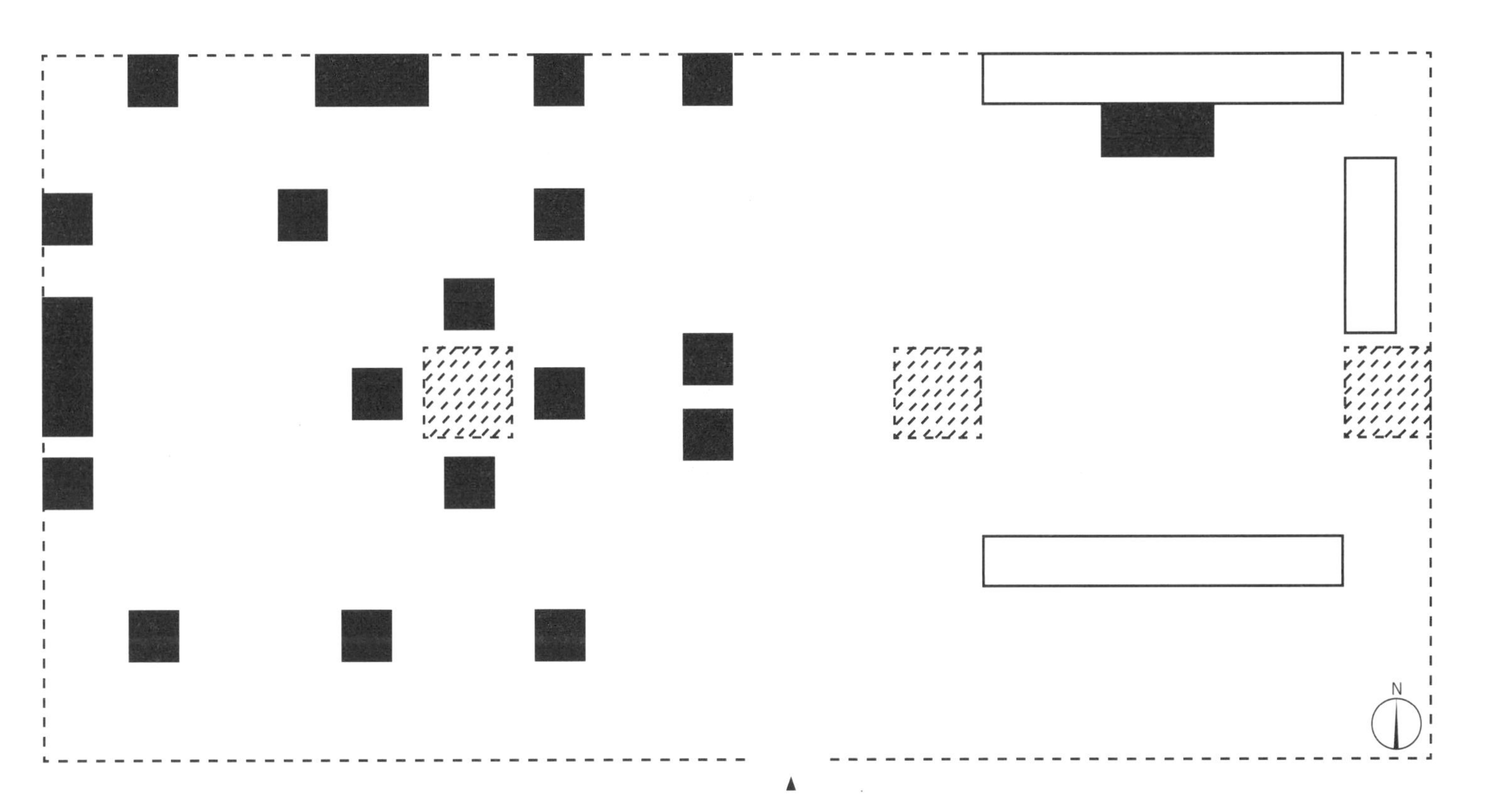

展品
Exhibits

展板
Display board

展场
Exhibition

柱
Pillar

六方形南官帽椅
ADA

四出头官帽椅

J I N G
P A R E D E S

榫卯
榫卯，是古代中国建筑、家具及其它器械的主要结构方式，是在两个构件上采用凹凸部位相结合的一种连接方式。凸出部分叫榫（或叫榫头），凹进部分叫卯（或叫榫眼、榫槽）。
攒边打槽装板（局部）
圆方结合裹腿
椅子后足穿过椅盘的结构
高束腰抱肩榫
走马销
方材角结合床围子攒接万字
平板明榫角结合
椅子后足穿过椅盘的结构
攒边打槽装板
方材角结合床围子攒接万字
柜子底枨高枨互让（各用大进小出榫）
椅子后足穿过椅盘的结构
厚板闷角榫结合
平板明榫角结合

Traditional culture inheritance and innovative design

传统文化是本民族自身对世界的一种认识，是人类群体的精神文明象征。如果失去了传统文化，就如同在黑暗中失去了烛火一样，失去了方向。世界是由多民族、多文化所组成的，相应产生了区别非常明显的多样的传统文化。传统文化的精神文明象征能够代表着人类群体，这也是一个民族区别于另一个民族最大的特征，延续和传承即是一个民族流传的目标。

新时代，交通将变得更便利，沟通将变得更频繁，类别多样的传统文化互相交织的发展面貌更广泛。更多的交流，更多的互通，使得大趋势趋于融合，不可避免地就是文化的趋同。一个民族的消亡往往和精神文明的消失挂上钩，于是保持本民族优秀的传统文化势在必行。

社会的发展日新月异，有的传统文化已经不再适应现代社会，如何保护和维持传统文化精华成为极大的挑战。一方面是要树立民族的自信心，相信民族带来的认同感和对未来的希冀；另一方面是要对本民族传统文化设计性保留，取其形，留其意，将新技术、新产业和谐融入。传统文化影响着我们的教育生活和国际交流，是一个由内向外扩散的过程，是以传统文化为思想核心，对外交流拓展融合的创新。传统文化传承下的设计创新不仅需要从孩童的教育内容入手，更要让孩童从心底上认识到民族精神文化的重要性，形成文化自信。带着这样的愿望，携手走入新时代。

Traditional culture is a cognition for the world by a nation, and it is a culture symbolization about the human spirit. A lost of traditional culture which likes lost in the dark without candlelight. The world combines a variety of nation and culture, and it creates some traditional culture with a noticeable difference. A human group can be recognized by their traditional culture. It is the most different from a nation to the other nation. A nation aims to continue and inherit its culture.

In the new age, traffic will become more natural, communication will become more frequent, and culture will be more extensive. Different cultures tend to similar to frequent communication and exchange. Extinction of a nation always relates to the disappearance of their spiritual civilization. Thus, protect their own excellent traditional cultures are no delay.

Some traditional cultures are no longer adapt to modern society with the rapid change of social development. There would be a big challenge in how to protect and keep the essence of traditional culture. Firstly, it should build up the confidence of the nation, believe the sense of identity from their nation and hope for the future. Besides, selective retention of traditional culture, such as Choose the shape and reserve the meaning. Also, integrating new technology and new industry with it. Traditional culture influences our education and international exchange, and it is a process from inside to outside. It is an innovation about expand and the mixture of international communication with the core of traditional culture thought. The innovation of design not only start from childhood education but also let children know the importance of national sprite and be confident in their culture. Bring this wish, go into the new world hand in hand.

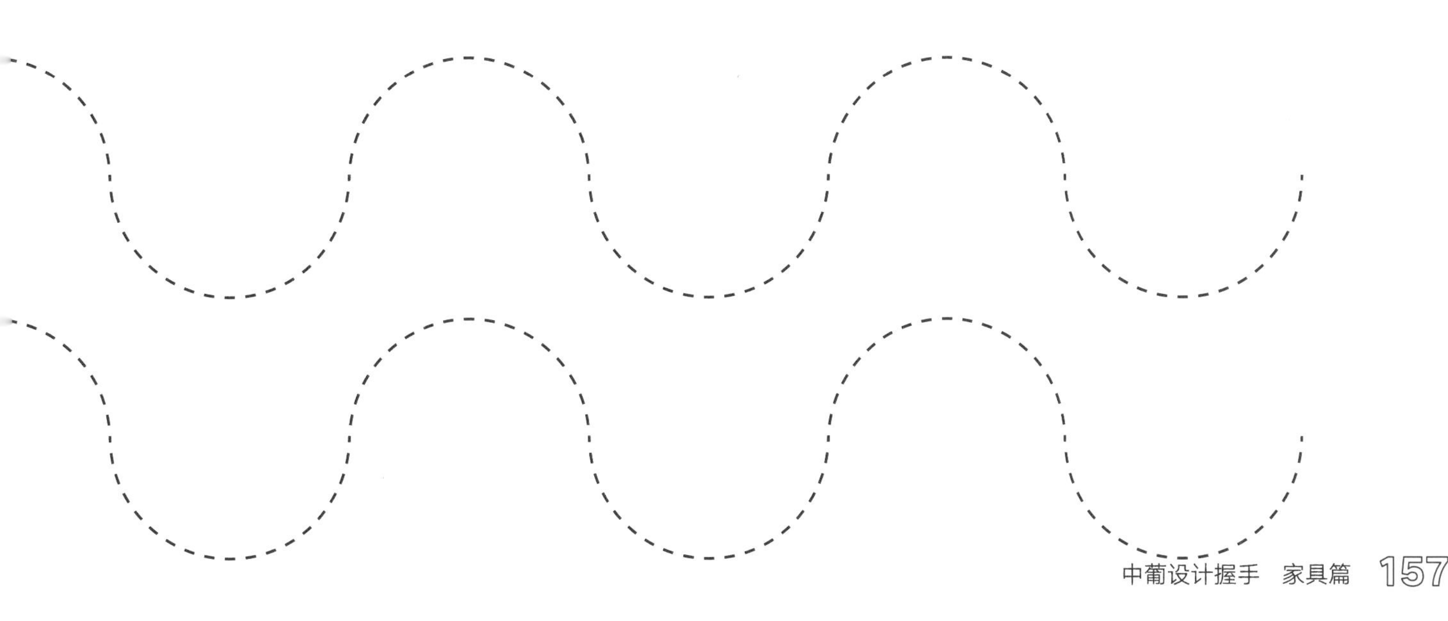

邀请函 Invitation

中葡设计握手 家具篇

Handshake between Chinese and Portuguese Design Furniture

从 / 北 / 京 / 到 / 帕 / 雷 / 德 / 斯

FROM BEIJING TO PAREDES

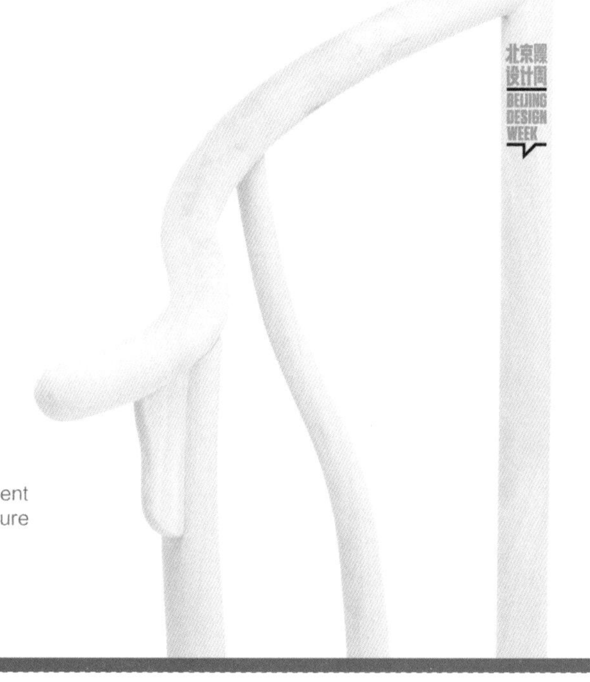

主办单位:
北京市人民政府外事办公室
北京建筑大学

Host:
Foreign Affairs Office of Beijing Municipal Government
Beijing University of Civil Engineering and Architecture

承办单位:
北京歌华设计有限公司

Organizer:
Beijing Gehua Design Company

B E I J I N G / P A R E D E S

Dear Sir/Madam:

The “Handshake between Chinese and Portuguese Design · Furniture” exhibition will open at september 9, 2016 (Friday) in Floor 13, Gehua Tower Block A. The favor of your presence is requested.

挚 邀

尊敬的　　　　先 生（女士）：

您好！兹定于2016年9月9日（星期五）10时在歌华大厦A座十三层举行“中葡设计握手·家具篇”展览开幕活动。

敬请光临为盼!

邀请函(190 x 190mm)
Invitation(190 x 190mm)

B E I J I N G / P A R E D E S

主办单位：
北京市人民政府外事办公室
北京建筑大学
承办单位：
北京歌华设计有限公司

Host:
Foreign Affairs Office of Beijing Municipal Government
Beijing University of Civil Engineering and Architecture
Organizer:
Beijing Gehua Design Company

中葡设计握手 家具篇

Handshake between Chinese and Portuguese Design Furniture

从 / 北 / 京 / 到 / 帕 / 雷 / 德 / 斯

FROM BEIJING TO PAREDES

2015「驻场计划」第二期

Phase II of Design-in-Residence Programmes

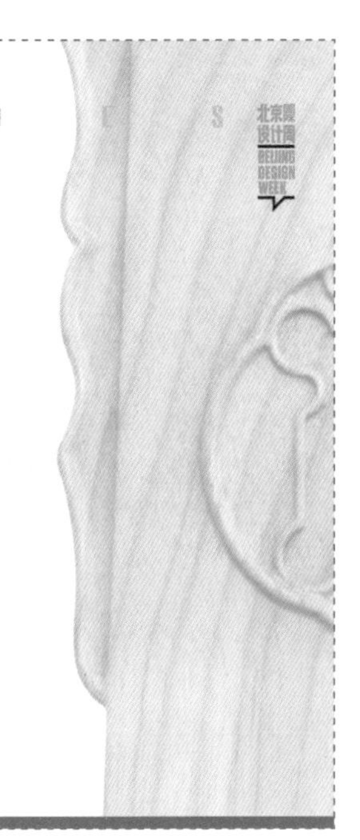

B E I J I N G / P A R E D E S

主办单位：
北京市人民政府外事办公室
北京建筑大学
承办单位：
北京歌华设计有限公司

Host:
Foreign Affairs Office of Beijing Municipal Government
Beijing University of Civil Engineering and Architecture
Organizer:
Beijing Gehua Design Company

中葡设计握手 家具篇

Handshake between Chinese and Portuguese Design Furniture

从 / 北 / 京 / 到 / 帕 / 雷 / 德 / 斯

FROM BEIJING TO PAREDES

2015「驻场计划」第二期

Phase II of Design-in-Residence Programmes

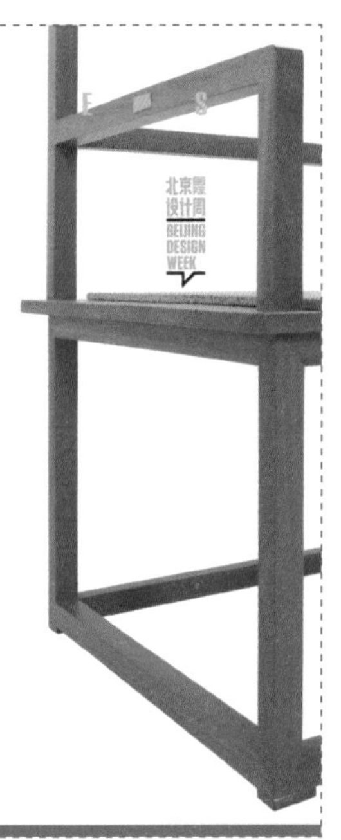

B E I J I N G / P A R E D E S

主办单位:
北京市人民政府外事办公室
北京建筑大学
承办单位:
北京歌华设计有限公司

Host:
Foreign Affairs Office of Beijing Municipal Government
Beijing University of Civil Engineering and Architecture
Organizer:
Beijing Gehua Design Company

中葡设计握手 家具篇
Handshake between Chinese and Portuguese Design Furniture
从/北/京/到/帕/雷/德/斯
FROM BEIJING TO PAREDES
2015「驻场计划」第二期
Phase II of Design-in-Residence Programmes

B E I J I N G / P A R E D E S

主办单位:
北京市人民政府外事办公室
北京建筑大学
承办单位:
北京歌华设计有限公司

Host:
Foreign Affairs Office of Beijing Municipal Government
Beijing University of Civil Engineering and Architecture
Organizer:
Beijing Gehua Design Company

中葡设计握手 家具篇
Handshake between Chinese and Portuguese Design Furniture
从/北/京/到/帕/雷/德/斯
FROM BEIJING TO PAREDES
2015「驻场计划」第二期
Phase II of Design-in-Residence Programmes

北京设计周
BEIJING DESIGN WEEK

大堂垂幅(6000 x 2600mm)
Hall curtain on the first floor(6000 x 2600mm)

B E I J I N G / P A R E D E S

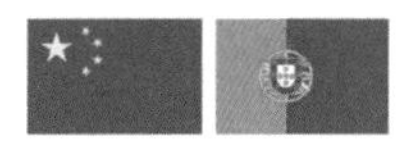

中葡设计握手 家具篇

Handshake between Chinese and Portuguese Design Furniture

从/北/京/到/帕/雷/德/斯

FROM BEIJING TO PAREDES

2015「驻场计划」第二期

Phase II of Design-in-Residence Programmes

主办单位:
北京市人民政府外事办公室
北京建筑大学

承办单位:
北京歌华设计有限公司

Host:
Foreign Affairs Office of Beijing Municipal Government
Beijing University of Civil Engineering and Architecture

Organizer:
Beijing Gehua Design Company

B E I J I N G / P A R E D E S

前言 Foreword

为促进中葡文化交流，北京市人民政府外事办公室联合北京歌华文化发展集团和葡中文化协会，于2015年6月-7月间共同举办了为期30天的第二期"中国青年设计师驻场四季计划"，主题为"家具设计"，地点在葡萄牙帕雷德斯市。

北京建筑大学等院校的10位师生设计师参与了本期文化交流活动。师生设计师根据对葡萄牙当地文化的理解和感悟，借助帕雷德斯市5家家具企业，协同设计制作出面向中欧消费者的当代家具。

"驻场计划"活动受到两国政府的高度重视，葡萄牙时任总统卡瓦科·席尔瓦还亲切接见了参加本季"驻场计划"师生。

作为双向交流活动，2016年9月，北京建筑大学将近年在"中国传统家具构法与制作技艺"研究中完成的8件明式家具"标准器"的复原作品与本期"驻场计划"中完成的10件当代家具作品在"中葡设计握手"展暨中外交流会上作对照展出，以呈现中葡两国家具文化面貌，启发人们对家具文化传承与创新的思考。

Under the theme of furniture design, the "Phase || of Design-in-Residence Programs" program was held between June and July 2015 in Parades Portugal. This program is launched by Foreign Affairs Office of Beijing Municipal Government, with the joint efforts of Beijing Gehua Cultural Development Group and China-Portugal Cultural Creative Cooperation Platform, to promote cultural exchange between China and Portugal.

Ten designers from Chinese universities such as Beijing University of Civil Engineering and Architecture joined this project. The designers added their understanding of Portuguese culture into their design works, and produced their design with the support of five furniture companies in Paredes. Their design works are for both Chinese and European consumers.

This program was highly valued by both governments. Mr. Anibal Cavaco Silva, President of Portugal, met these young designers.

As a two-way exchange event for inspiring the thinking on innovation and inheritance of traditional furniture, people will see both Chinese and Portuguese culture in the furniture exhibition & exchange event of "Handshake between Chinese and Portuguese Design.Furniture" this September. Beijing University of Civil Engineering and Architecture will put on display ten modern furniture design works made during the "Design-in-Residence Programs", and eight Ming-style furniture reproductions made during their research on traditional furniture structure methods & craftsmanship.

B E I J I N G

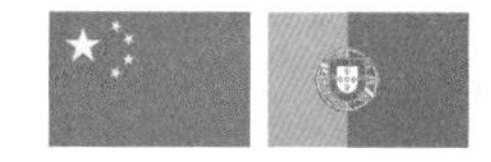

北京国际设计周
BEIJING DESIGN WEEK

中葡设计握手 家具篇
Handshake between Chinese and Portuguese Design Furniture

从/北/京/到/帕/雷/德/斯
FROM BEIJING TO PAREDES
2015「驻场计划」第二期
Phase II of Design-in-Residence Programmes

参观由此向前

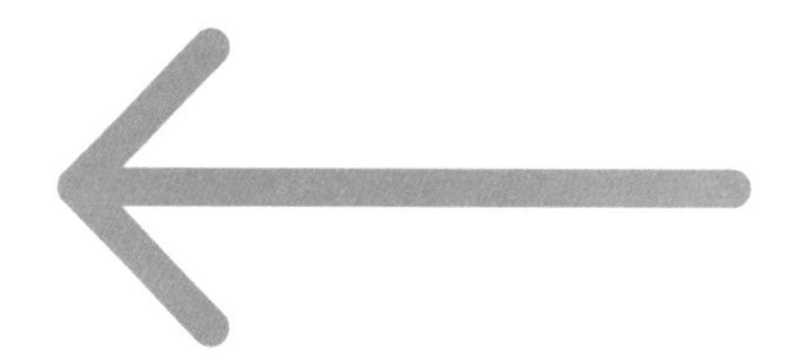

P A R E D E S

前言主背板(6000 x 3600mm)
Foreword background(6000 x 3600mm)

导向标识牌(100 x 200mm)
Guide sign(100 x 200mm)

榫卯

榫卯，是古代中国建筑、家具及其它器械的主要结构方式，是在两个构件上采用凹凸部位相结合的一种连接方式。凸出部分叫榫（或叫榫头），凹进部分叫卯（或叫榫眼、榫槽）。

三根直材交叉

攒边打槽装板（局部）

圆方结合裹腿

椅子后足穿过椅盘的结构

圆香几攒边打槽

圆柱二维丁字结合榫

高束腰抱肩榫

走马销

方材角结合床围子攒接万字

扇形插肩榫

平板明榫角结合

椅子后足穿过椅盘的结构

插肩榫变体

攒边打槽装板

方材角结合床围子攒接万字

柜子底枨两枨互让（各用大进小出榫）

椅子后足穿过椅盘的结构

加云子无束腰裹腿杌凳腿足与凳面结合

厚板闷角榫结合

平板明榫角结合

榫卯展板背面(3600 x 3600mm)
The back of the mortise and tenon board(3600 x 3600mm)

B E I J I N G / P A R E D E S

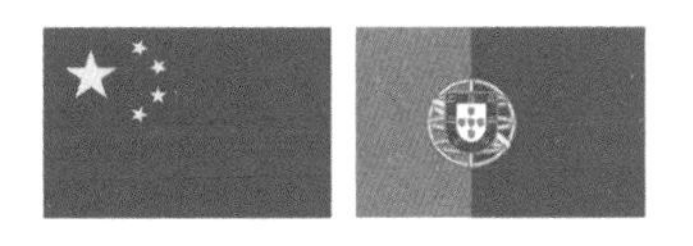

北京国际设计周 BEIJING DESIGN WEEK

中葡设计握手 家具篇

Handshake between Chinese and Portuguese Design Furniture

从/北/京/到/帕/雷/德/斯

FROM BEIJING TO PAREDES

2015「驻场计划」第二期

Phase II of Design-in-Residence Programmes

主办单位： **Host:**

北京市人民政府外事办公室 Foreign Affairs Office of Beijing Municipal Government

北京建筑大学 Beijing University of Civil Engineering and Architecture

承办单位： **Organizer:**

北京歌华设计有限公司 Beijing Gehua Design Company

榫卯展板正面(3600 x 3600mm)

The front of the mortise and tenon board(3600 x 3600mm)

BEIJING

四出头官帽椅

Armchair with four protruding ends

ADA 家具构法与制作技艺研究所

复原者 Replicator:

北京建筑大学 - 工业设计 111 班
指导教师：郎世奇 陈静勇 尚华

Industrial Design Class of 111
Advisory teachers: Lang Shiqi, Chen Jingyong, Shang Hua

规格 Dimension:

长 × 宽 × 高　589×493×1127（mm）

此椅原型为“明式四出头官帽椅”，是我国明式家具中椅子造型的一种典型款式。其造型像古代官员的帽子而得名，上下无一丝装饰，结构简练之极，完全采用线条和弧度来处理，线条曲直相间，方中带圆，充分体现了明式家具简洁明快的特点。

The prototype of the armchair is "official's hat armchair with four protruding ends of Ming Style". It is a very classic chair among Ming-style furniture. The appearance of the chair is like official's hat in ancient times. It is concise and simple, with no decoration. The curve and straight lines of the chair fully expresses the simple style of Ming furniture.

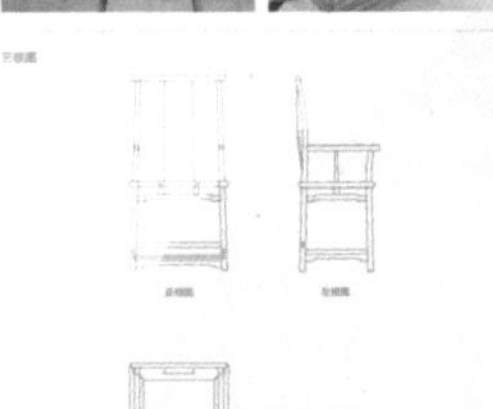

PAREDES

BEIJING

灯挂椅

Lamp-hanger chair

ADA 家具构法与制作技艺研究所

复原者 Replicator:

北京建筑大学 - 工业设计 091 班
指导教师：陈静勇 郎世奇

Industrial Design Class of 091
Advisory teachers: Chen Jingyong, Lang Shiqi

规格 Dimension:

长 × 宽 × 高　490×428×1090（mm）

此椅原型为“明式灯挂椅”，灯挂椅是一种历史悠久的汉族椅式家具，因其造型好似南方挂在灶壁上用以承托油灯灯盏的竹制灯挂而得名。整体感觉是挺拔向上，简洁清秀，为明式家具的代表作。

Its prototype is the Ming-style lamp-hanger chair. The lamp-hanger chair is very ancient furniture for Han people. It is tall and straight, simple and elegant. The lamp-hanger chair is a magnum opus among Ming-style furniture.

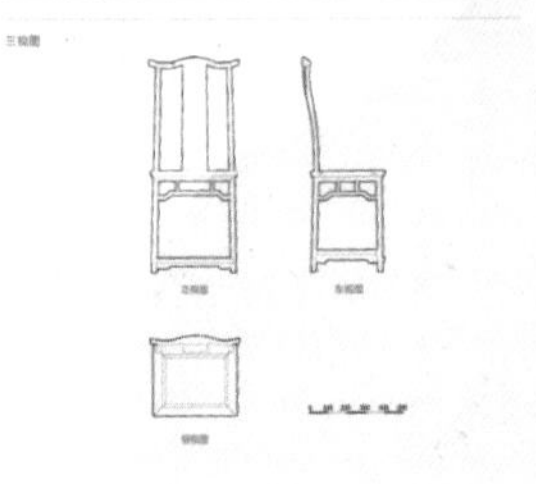

PAREDES

BEIJING

圈椅

Round-backed armchair

ADA 家具构法与制作技艺研究所

复原者 Replicator:

北京建筑大学 - 工业设计 111 班
指导教师：陈静勇 郎世奇 韩风 尚华

Industrial Design Class of 111
Advisory teachers: Chen Jingyong, Lang Shiqi, Han Feng, Shang Hua

规格 Dimension:

长 × 宽 × 高　704×605×1022（mm）

此椅原型为“明式素圈椅”，大体光素，椅盘以上为圆材，以下外圆内方。三面素牙条，除靠背板有浮雕、扶手下有小角牙外，可以视作圈椅的基本形式。

The prototype of the chair is "round-backed armchair of Ming style". It is simple and neat. It is supported with spandrel in three sides. Apart from the reliefs on the back and the little spandrels under the arms, it shows the basic type of round-backed armchair.

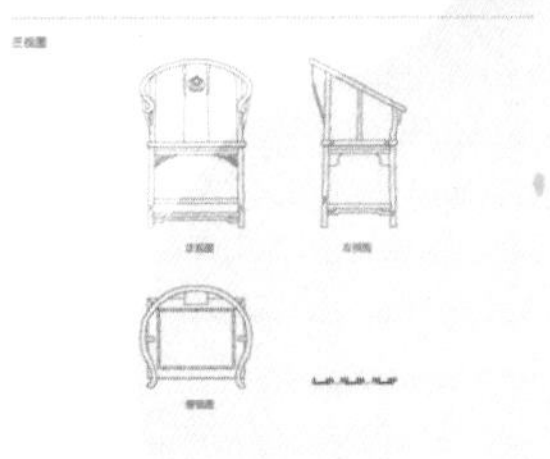

PAREDES

BEIJING

六方形南官帽椅

Southern official's hat armchair

ADA 家具构法与制作技艺研究所

复原者 Replicator:

北京建筑大学 - 工业设计 121 班
指导教师：陈静勇 郎世奇

Environment Design Class of 121
Advisory teachers: Chen Jingyong, Lang Shiqi

规格 Dimension:

长 × 宽 × 高　800×563×962（mm）

此椅原型为“明式六方形南官帽椅”，六方椅在明式家具中极为罕见，此椅尺寸甚至大于一般扶手椅，采用了较为复杂的线脚，是一个大胆的创新。此椅造型独特，虽是变体，但意趣清新，自然大方，无矫揉造作之感。

The prototype of the chair is hexagon-shaped south officer's hat armchair with Ming style. It is very rare among Ming-style furniture. It is larger than normal armchairs. Its sophisticated skintle is a bold innovative design. Though it is a variant of the original one, it is novel yet natural.

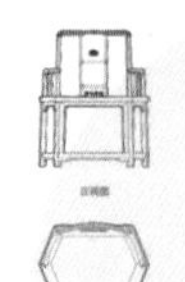

PAREDES

BEIJING

玫瑰椅

Rose chair

ADA 家具构法与制作技艺研究所

复原者 Replicator:

北京建筑大学 - 工业设计 121 班
指导教师：陈静勇 郎世奇 尚华

Environment Design Class of 121
Advisory teachers: Chen Jingyong, Lang Shiqi, Shang Hua

规格 Dimension:

长 × 宽 × 高　612×426×828（mm）

此椅原型为“明式券口靠背玫瑰椅”。玫瑰椅是明式扶手椅中常见的形式，其特点是靠背、扶手和椅面垂直相交，尺寸不大，用材较细，予人一种轻便灵巧的感觉，为各种椅子中较小的一种，用材单细，造型小巧美观。

The rose chair is a very normal Ming-style armchair. Its feature is that the back, the arm and the seat are in vertical with each other. With proper size and delicate material, it shows handy, smart and elegant fashion.

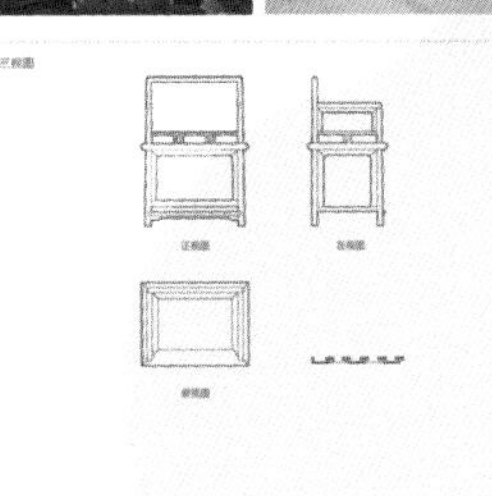

PAREDES

BEIJING

一腿三牙罗锅枨方桌

Three spandrels humpbacked stretcher Square table

ADA 家具构法与制作技艺研究所

复原者 Replicator:

北京建筑大学 - 工业设计 121 班
指导教师：陈静勇 郎世奇

Industrial Design Class of 121
Advisory teachers: Chen Jingyong, Lang Shiqi

规格 Dimension:

长 × 宽 × 高　980×980×880（mm）

此椅原型为“明式黄花梨一腿三牙罗锅枨方桌”。此种方桌是明式方桌中的一种常见形式。腿足及罗锅枨上犀利有力的锐棱峭拔精神，使方桌显得骨相清齐，劲挺不凡。

Its prototype is the Ming-style Huanghuali-wood three spandrels humpbacked stretcher square table, which is very common in Ming-style square table. The sharp and brisk lines on the leg and the hump-baked stretcher give the table a refreching and strong spirit.

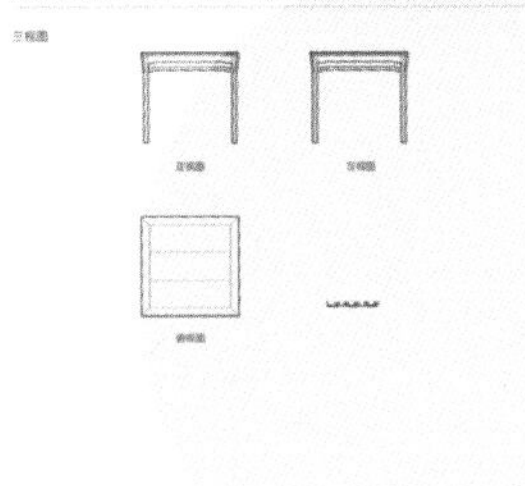

PAREDES

BEIJING

夹头榫带屉板平头案

Elongated bridle joint flat-top recessed-leg table with drawers

ADA 家具构法与制作技艺研究所

复原者 Replicator:

北京建筑大学 - 工业设计 101 班
指导教师：陈静勇 郎世奇

Industrial Design Class of 101
Advisory teachers:Chen Jingyong, Lang Shiqi

规格 Dimension:

长 × 宽 × 高　707×335×790（mm）

此椅原型为“明式夹头榫带屉板平头案”。此带屉板平头案为典型平头案的变体。为了增加条案的使用空间与面积，案面下有一屉板，镶入腿足间枨子的槽口，枨子作肩接入腿足。

The prototype of this work is Ming-style Elongated bridle joint flat-top recessed-leg table with drawers. This work is a variant of classic flat-top table. To increase the space for use, there is the drawer under the table top. It is edged into the notched stretcher The stretcher, as a shoulder, is then attached to the leg.

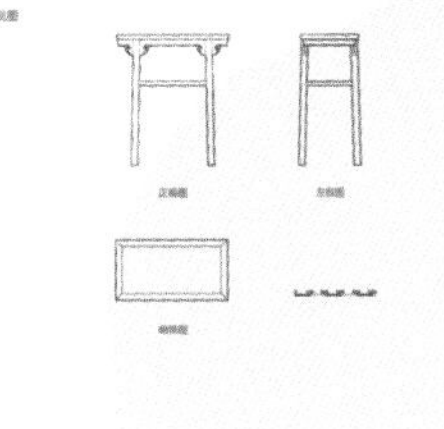

PAREDES

BEIJING

翘头案

Head desk

ADA 家具构法与制作技艺研究所

复原者 Replicator:

北京建筑大学 - 环境设计 131 班
指导教师：陈静勇 郎世奇 韩风 尚华

Environment Design Class of 131
Advisory teachers: Chen Jingyong, Lang Shiqi,
Han Feng, Shang Hua

规格 Dimension:

长 × 宽 × 高　1400×463×836（mm）

此案原型为"明式夹头榫带托子翘头案"。翘头案案面两端上翘。俗称飞角。翘头增加了案子的雅致轻盈，不失尚礼文秀。在中国古代常置于厅堂，造型灵巧庄重，是屋内器物陈设装饰的重要承具。

The prototype of the table is " Ming-style Elongated bridle joint everted flanges Recessed-leg table with Tray." Both ends of the table are up. It is commonly known as flying horn. Everted flanges increase the sense of elegance and lightness of table, and not lost gentle and pretty. It always set in the hall in ancient China. It is a significant table which shape is dexterity and solemnity with objects placed in the house.

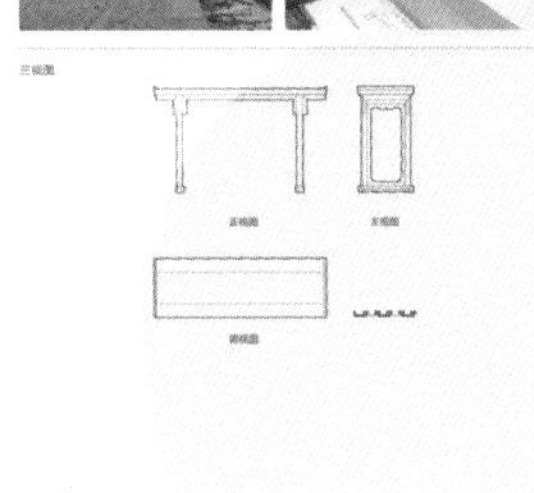

PAREDES

八件明式家具展布(1000 x 5000mm)
Eight state of Ming-style furniture of hardwood(1000 x 5000mm)

BEIJING

总统桌 President·table

杨 琳 Lynn

北京建筑大学建筑学院设计学系副教授。

Associate Professor of Design at Beijing University of Civil Engineering and Architecture.

设计理念 Design Idea

此总统桌的设计诠释了“中国设计葡国工艺”，为纪念总统会见“驻场计划”北京设计师代表团而设计。由桌面、桌腿、托泥三个部位组成，部件间通过螺栓连接，形成梯形框架形态；采用当地胡桃木制作，极简的线条体现结构和材质的美感。

This table embodies Chinese design and Portugal crafts. It is designed to mark Portugal President's meeting with Beijing designers in the "Artist-In-Residence" program. This design is composed of the table top, legs and leg supports. It is made of local walnut wood. The neat lines show its prominent structure and material.

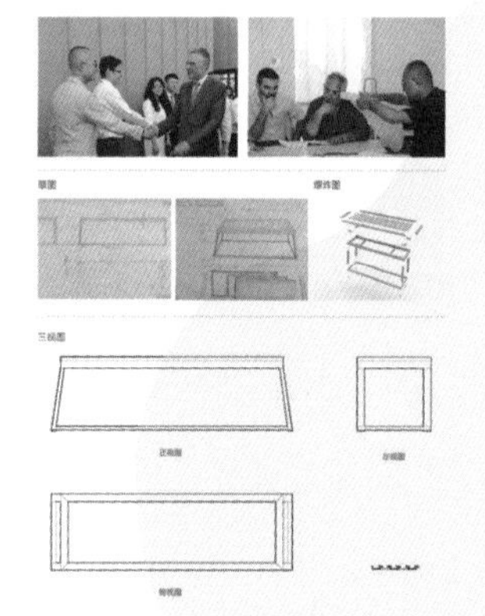

PAREDES

BEIJING

总统椅 President·chair

邹 乐 Mark

北京建筑大学设计学专业 2014 级硕士研究生。

Graduate student(enrolled in 2014) in design of Beijing University of Civil Engineering and Architecture.

设计理念 Design Idea

总统椅的设计诠释了“中国设计葡国工艺”，为纪念总统会见“驻场计划”北京设计师代表团而设计。极简的线条体现结构和材质的美感，采用胡桃木制做，椅面铺有软木；作品由椅腿、坐凳和椅背扶手 3 个独立部位组成，通过螺栓连接，方便独立自主拼装、运输。

This chair embodies Chinese design and Portugal crafts. It is designed to mark Portugal President's meeting with Beijing designers in the "Artist-In-Residence" program. The main structure of the chair is made of walnut wood. Its seat is covered with cork wood. This design is composed of the legs, the seat and the back and arm. They are connected with bolts, convenient for assembling and transportation.

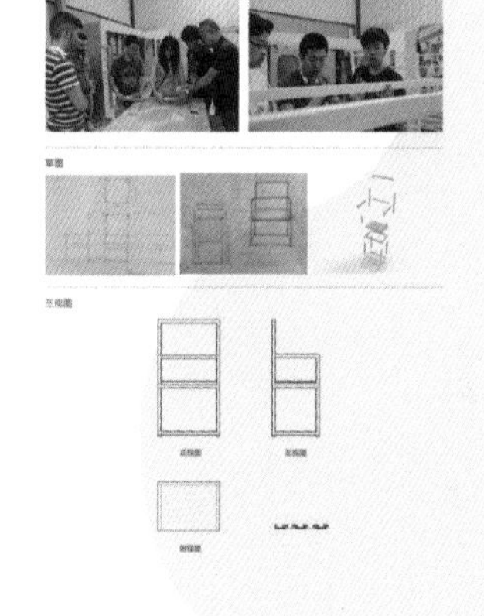

PAREDES

BEIJING

咖啡几 Coffee table

朱宁克 Nick

北京建筑大学设计学系副主任，硕士学位，主要从事室内、产品、家具设计教学与研究。

Vice Dean of Design in Beijing University of Civil Engineering and Architecture with a master's degree. Teaching and studying of interior, production and furniture design.

设计理念 Design Idea

圆形咖啡几，造型简洁。四条木质桌腿穿过桌面形成了 4 个木质“犄角”，增加了作品的趣味性。

This is a round coffee table with simple style. The four wooden legs protrude through the table top, making it fun.

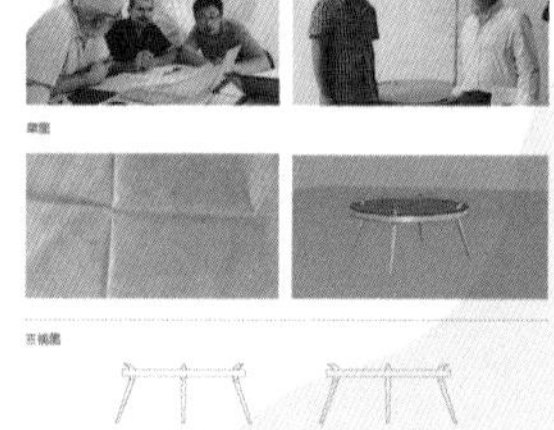

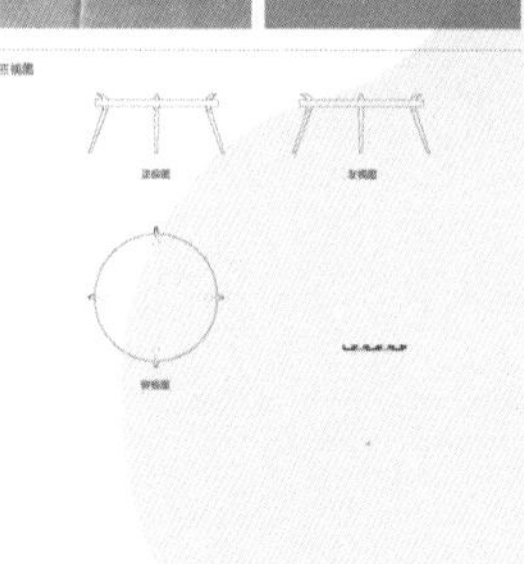

PAREDES

BEIJING

缠绕 Binding

韩 风 Kevin

北京建筑大学设计学系讲师，博士学位，主要从事室内和家具设计。

Lecturer of design of Beijing University of Civil Engineering and Architecture Interior and furniture designer.

设计理念 Design Idea

圆几设计围绕日常家用需求展开，以一条缠绕的折线为形态创意，从抽屉拉手开始，与腿足和装饰线形成连贯折线，最终止于几面。抽屉内部隔板分布合理，保证日常功能使用。

The round table designs for meeting daily household needs. The inspiration is a binding line starting from the drawer handle, the legs, to the table top. Rational layout of the partition in the drawer makes sound daily use.

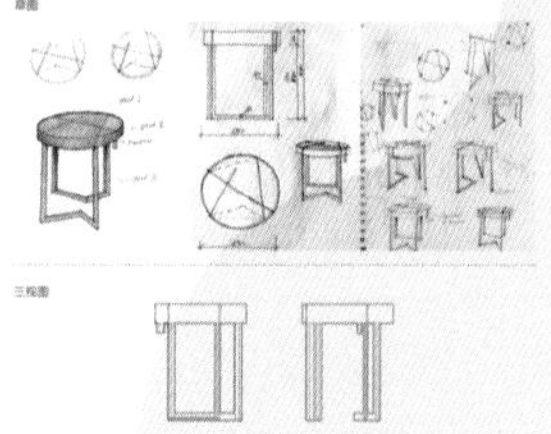

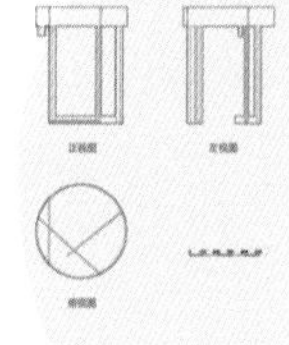

PAREDES

B E I J I N G

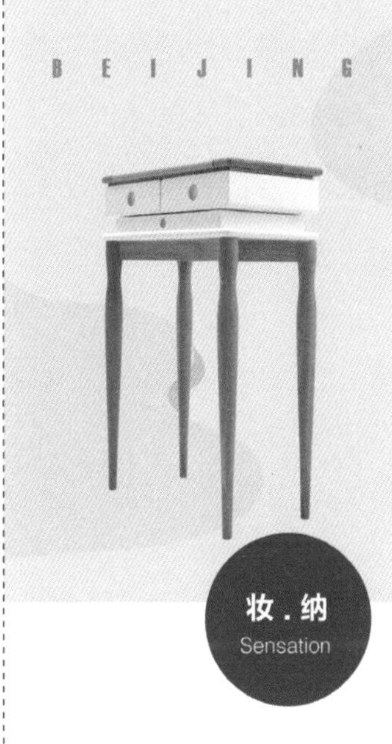

妆 . 纳
Sensation

孙 滨 Sunny

北京朝阳实验小学高级教师，曾获北京市朝阳区优秀青年教师称号。

Senior Teacher of Beijing Chaoyang Experimental Primary School, Winner of the Title "Outstanding Young Teacher of Chaoyang District"

设计理念 Design Idea

梳妆桌表面看起来像一张开心的笑脸，赋予此作品随性、开朗的性格。造型简洁、占用空间小，揭开桌面是一面镜子，桌体设上下两层，可分层存放彩妆、首饰，适用于年轻女性。

The surface of the dresser is a smiling face, giving a casual and cheerful feeling. The design is simple and space-saving. You will see a mirror when uncovering the table top. There are two selves inside for keeping cosmetics and jewelry.

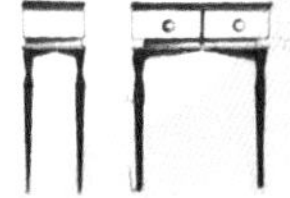

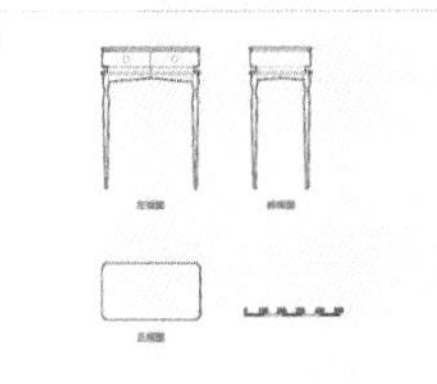

P A R E D E S

B E I J I N G

分 . 格空间
Separate space

闫卓远 Joy

北京建筑大学设计学专业 2014 级硕士研究生；本科毕业于沈阳工业大学工业设计专业。主要研究方向为空间环境与产品设计。

Graduate student (enrolled in 2014) in design of Beijing University of Civil Engineering and Architecture. Bachelor's degree in design from Shenyang University of Technology. Research area is space environment and product design.

设计理念 Design Idea

此置物架的设计思路寻求个性化需求与批量生产相结合，以适应不同空间环境为目的。在相同单元的不同组合中，人们融入自己的创意，在规则与自由、理性与感性之中探索独特的表达方式。

This shelf pursues both individual needs and mass production, and aims to fit different space. Through different combination, users can add their creativity into the same space, thus achieving unique expression between rules and freedom, sense and sensibility.

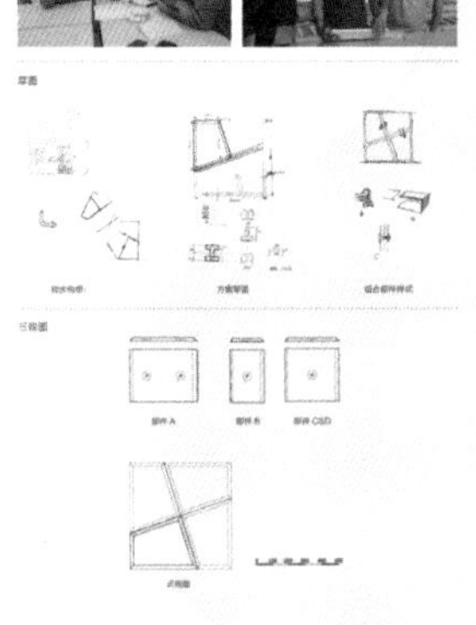

P A R E D E S

B E I J I N G

清风
Brezze

赵安路 Zhao Anlu

北京建筑大学工业设计专业 2012 级本科生。

Undergraduate student (enrolled in 2012) in design of Beijing University of Civil Engineering and Architecture.

设计理念 Design Idea

作品灵感来源于栏栅，横纵结合，穿插有序，通过最简约直线条构成产品外观和结构，主要用作鞋柜，上层抽屉可存放小杂物，镂空门扇便于通风，配以鞋凳方便使用。

The design is inspired by fences. The appearance and structure of this work is formed with neat lines. It can be used as a shoe cabinet. Its upper drawer is for keeping small things and the hollow-out door for letting air in. It would be more convenient when used together with a stool.

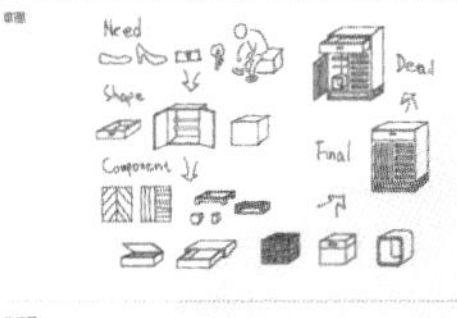

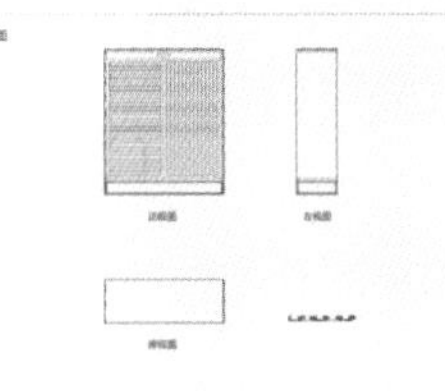

P A R E D E S

B E I J I N G

一 方 凳
A cube

沈相宜 Shen Xiangyi

北京建筑大学工业设计专业 2012 级本科生。

Undergraduate student (enrolled in 2012) in design of Beijing University of Civil Engineering and Architecture.

设计理念 Design Idea

坐凳设计灵感来源于一块积木。色彩鲜亮的小方块凸显出柔软舒适的坐面，底部支撑结构则增强了坐具的稳定性。

It is inspired by a piece of wood block. The bright-colored block highlights the comfort and softness of the seat and the bottom provides stable support.

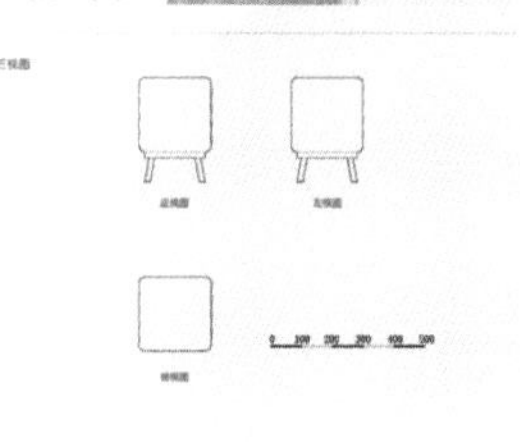

P A R E D E S

BEIJING

融错 . 空间

The interlaced dimensions

赵雄韬 Zhao Xiongtao

北京朝阳实验小学高级教师，曾获北京市朝阳区优秀青年教师称号。

Senior Teacher of Beijing Chaoyang Experimental Primary School, Winner of the Title "Outstanding Young Teacher of Chaoyang District"

设计理念 Design Idea

衣架结构清晰，一方一圆，一纵一横，两者交错，空间融合，使用便捷，是忙碌的城市生活中的一个简单的角落。

The clothes stand has clear structure. It keeps balance between square and round, vertical and horizontal elements. It provides a user-friendly space in busy city life.

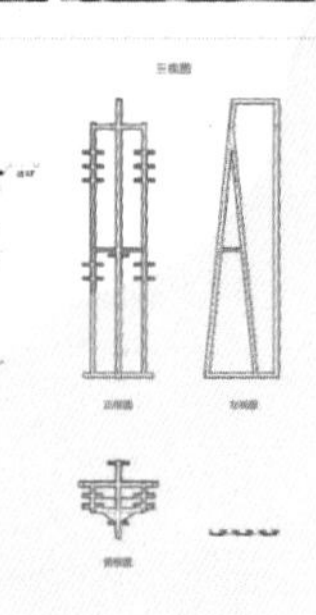

PAREDES

BEIJING

小木马

The little horse

郗鑫鑫 XI xinxin

北京建筑大学设计学专业2015级硕士研究生。研究方向为空间环境与产品设计。

Graduate student (enrolled in 2015) in design of Beijing University of Civil Engineering and Architecture, Research area is space environment and product design.

设计理念 Design Idea

小木马儿童椅采用葡萄牙当地木材，既考虑儿童使用的趣味性，也关注使用安全问题。可通过零件的拆装适应儿童不同年龄段的可持续使用需求。

Using Portugal local wood, the Little Horse takes into consideration child interest and safety in use. Sustainable use for children with different age can be realized through dismantling or installing components.

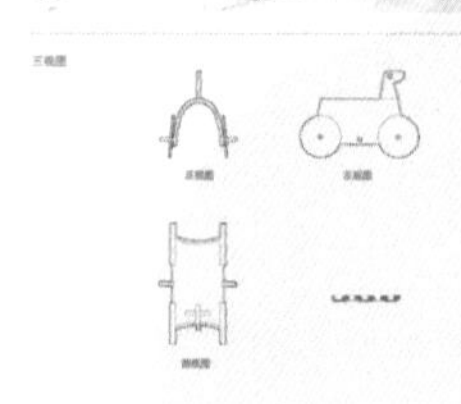

PAREDES

10件青年设计师家具设计展布(1000 x 5000mm)

Ten states of the furniture of young designers(1000 x 5000mm)

歌华大厦一层大厅正在挂展布
The workers on the first floor are hanging state in Gehua building

歌华大厦十三层展厅正在布展
The 13th-floor exhibition hall of Gehua building was being spread

RETHINK

悟

交流

杨琳

在第二期“中国青年设计师驻场四季计划”的整个过程中，我作为中方设计师团队的领队教师，与葡方波尔图大学的教授们一起指导每一个队员的家具设计。在设计过程中，我认为最重要的不是草图、模型，而是“交流”，这种交流不只是于葡方设计师、教授之间，更有对当地地域、人文、区域特点的认知也是必不可少的一部分。作为领队教师，此次交流活动对于我来讲如何引导学生比如何教导学生更加重要。所以每一次参观、每一次考察我都鼓励学生多去探寻、多去观察、多去记录，只有亲身经历过，这些隐形的设计因素才能根植于他们的思想中，才能在设计当中充分发挥与应用。

在整体的设计过程中，葡方教授与中方设计师都是以交流会的形式进行沟通。虽然在语言上我们之间有不相通的地方，但是在设计方案时有些认知是相同的，在交流时经常会产生共鸣，这也是我有所感触的部分。除了设计方面外，工厂的现场参观也是我认为比较有意义的部分，能够充分了解到欧洲的家具制造业目前的整体水平，也让学生们了解到从设计到图纸再到工厂大样的整体流程，同时也增强了学生们与工厂的沟通合作。通过双方的交流，对设计方案不断完善、不断优化进而做出自身满意的作品。在此次设计交流的最后，每一位师生的作品都参加了帕雷德斯市当地的展览，这也是我们此次交流过程中表达“中国设计”的媒介，让世界了解中国设计师的设计作品，这是值得骄傲的事情。

Communication

Yang Lin

In the whole process of "Phase || of Design-in-Residence Programs of Chinese Young Designers in Four Seasons," as a leader of the Chinese designer team, I worked with the Portuguese Porto University professors to guide each team member's furniture design.In the design process, I found the most important thing was not to sketch, model, but "to communicate".This communication was not only between Portuguese designers, professors, but also a kind of cognition on the local geographical, cultural, and regional characteristics. This cognition was an essential part. As a lead teacher, the exchange activities were more important for me to guide students than to teach them. So every visit, I always encourage students to explore more than once, more to observe, more to record. Only through personal experience, these invisible design factors can be rooted in their thoughts, and fully utilized and applied in design.

In the overall design process, Portuguese professor and Chinese designer communicated in the form of exchange meeting. Although our language is different, we share the same recognition and thoughts of design, some of them are the same in designing a program, which is of significant crucial in our communication. In addition to the design, the factory site visit is also a more meaningful part, which can not only make us fully understand the current overall level of European furniture manufacturing industry and the whole process of design from the drawings to the factory bulk sample, but improve the communication with factories as well. Through the exchange between the two sides, the design scheme is continuously enhanced and continuously optimized to do its satisfactory works. At the end of this design exchange, workpieces of every teacher and student are on display in the local exhibition in Paredes. This exchange is the medium for expressing Chinese Design" so that the world can make a full understanding of Chinese designing workpiece, which is a matter of pride.

波尔图
Porto

文化解读与文化融合

邹乐

感叹时光如流，转眼间葡国之行已经是两年多之前的事情了，现在想来往事却也依旧历历在目。在葡萄牙所邂逅的一切人、事物以及场景都根植在我的记忆中，作为我至今都记忆犹新的人生片段。

葡国之行对于我个人来讲与其说是设计交流的旅程，不如说它是一次文化解读的旅程。我认为，无论是建筑也好、家具也罢，都是各国人民在经历历史变迁过程中文化演进的产物。所以对于设计师来说，想要设计好一件“产品”（这里指的是广义产品），一定要切身去交流、去触碰、去品尝、去聆听、去感受，充分从感知方面去交融，才能进行深刻地文化剖析，才有可能会设计出好的作品。此次在葡萄牙所经历的一段，都为我设计“总统椅”作品打下了文化层面的基础，至于设计基础，纯粹是我个人表现文化理解的一种介质、方法了。

在设计方面，此次我设计时的主要难点在于文化的融合。作品主要是为纪念葡萄牙时任总统接见双方设计团队这一事件而设计的。在这把椅子的设计中结合了两国的文化，孕育出符合主题的设计作品。所以我在设计中融入对中葡两国文化的理解，对传统家具类型的认识，以及呈现的仪式感。

总而言之，此次活动对我来说是一次交流历练的机会。深度体验的模式也为日后中葡设计交流工作的展开打下了一定的基础。

Cultural interpretation and integration

Zou Le

Time flow, an instant trip to Portugal was a matter two years ago, but now I still fresh in the past and it always vivid. All the people, things and scenes in Portugal were rooted in my memory, as a fragment of life that I still remember.

It is better for me to say that it is a journey of cultural interpretation rather than a journey of design communication for me. I think that no matter whether it is architecture or furniture, it is the culture of people of all countries in the process of historical changes. For designers, if they want to design a good product and conduct a profound cultural analysis, they must experience the exchange, the touching, the tasting, the listening, the feeling and thoroughly blending from the perspective of perception. Designing a good product is possible. All I experienced that was the cultural foundation for the design of the "President's Chair" in Portugal. The basement of the plan was merely a personal expression of cultural understanding.

Concerning design, the main difficulty in my plan is the integration of culture. The work is mainly designed to commemorate the event which was the design team of both parties met by the president of Portugal. The design of this chair combines the cultures of both countries to create a theme-like design. Therefore, I incorporated in the design of the understanding of the culture of both counties and the understanding of the type of traditional furniture to present the sense of ceremony.

All in all, this event is a chance for me to exchange experiences. In-depth experience of the model for the future development of China-Portugal design and communication laid a solid foundation.

方案讨论中
Discussion

方案讨论中
Discussion

收获

朱宁克

2015年6月27日，我们10位中国设计师从北京启程赴葡萄牙。在接下来的一个月内，团队入驻在葡萄牙帕雷德斯市设计孵化器，在此住宿及工作。葡方安排了葡萄牙阿威罗大学设计学院5名教师协同中方团队共同设计，和当地5家家居制造企业配合设计师进行方案深化及样品制作。一个月的共同设计期间，中葡双方设计师共同讨论方案，参观工厂、当地博物馆、展览馆、波尔图大学等，并与当地生产企业协调的生产事宜。一年后，当时完成的10件设计作品参加了“2016北京国际设计周”的展出，中葡双方嘉宾及多家媒体进行了参观、报道，整个活动成果丰硕，实现了两国设计文化的“双向交流”。

记得双方设计师在确定设计对象的过程中就面临了来自不同的文化背景的巨大冲撞。中方设计师对产品的出发点多从感性出发，“葡萄牙印象”作为我的设计作品的出发点，是我们常采用的方式。但葡方设计师对此却不以为然，他不停地问这个家具是给谁用的，多高，会不会贵？等等……

太多的不同观点出现在整个活动中，给我们每个人带来的有不解、痛苦，也有成果、欢乐。也许正这是我们的收获。

Harvest

Zhu Ningke

On June 27, 2015, 10 of our Chinese designers departed from Beijing to Portugal. In the next month, the team settled in the design incubator of Paredes, Portugal, where we stayed and worked. The Portuguese people arranged five teachers from the Faculty of Design of Aveiro University to work together with the Chinese team, and they cooperated with five furniture manufacturing enterprises to deepen the sketch and making samples. During the one-month co-design period, Portuguese designer discussed the program with Chinese designers and they visited factories, local museums, exhibition halls and the University of Porto, then they coordinated production with the local manufacturing enterprises. A year later, ten design works which completed at that time took part in the exhibition of "Beijing Design Week 2016". The leaders and guests from China, Portugal, and many media visited and reported on it. The whole activity has achieved fruitful results and has realized the "two-way exchange" of design culture between the two countries.

In the process of determining the design object, the designers of both sides faced a massive collision from different cultural backgrounds. The Chinese designer's starting point for the product is from the perceptual, "Impression of Portugal" as the starting point of my design work is the way we often use. However, Portuguese designers do not agree with this, and he kept asking the furniture make for whom? How high? Will it be expensive? And many more.

Too many different views appeared in the whole activity, giving each of us a lot of incomprehension, confusion, joy and fruitful results. Maybe that's what we've got.

“驻场计划”中的葡萄牙朋友们

韩风

对于我们而言，葡萄牙“驻场计划”之行的丰富精彩在于每一处风景、每一餐美味、每一笔勾画、每一件家具，而这每一个瞬间都离不开我们这些活跃的主角。正是这些呈现在我们身边的片段，构成了一段段千姿百态的葡萄牙“驻场计划”故事。

这个故事里有我们这群“驻场计划”中国设计师，也少不了葡萄牙的设计师朋友们。

丰富多彩的葡萄牙生活

我们不远千里的赶来葡萄牙参加“驻场计划”家具设计专题，自然与在国内有所不同。而要有所不同，最大的挑战在于如何在如此短暂的时间里找到中葡设计握手的文化结合点。

从很多像帕雷德斯一样的小城镇，到波尔图、里斯本等大城市，再到水乡阿威罗、古都吉马良斯等历史名城一路走来，从绘画、雕塑、青花瓷、服装，到家具的形态、纹样、工艺和细节，在葡萄牙“驻场计划”相关老师和朋友们的陪同下，我们每人都通过不同的视角了解到这个陌生国度的历史和文化。

我们从早市上了解到小镇人的日常细节，从三餐中体味着当地人的饮食习惯，从切身参与到葡萄牙国庆等几个重要节日中感受到了葡萄牙生活节奏与文化，从每天的讨论和学习中了解到葡萄牙人的设计思维。正是点滴的细节都给予我们很大的启发，每个人的作品都经历了不断的自我否定与优化完善的过程。

“驻场计划”中的葡萄牙设计师

要说身边这几个“驻场计划”葡萄牙设计师最大的外貌特征，那一定非他们那一脸标志性的络腮胡莫属了，成熟稳重而设计师范儿爆棚。他们性格相对随和，每天和他们探讨设计、交流都很放松，很容易就拉近彼此的距离。他们表情和肢体语言极其丰富，说话的时候眉飞色舞，似乎每一个字都要配合一定的手势和表情。

几名葡萄牙设计师的敬业精神着实令人钦佩，他们不会放过任何细节，很多次与我们的讨论都要持续七八个小时或直到半夜。他们从欧洲人的生活习惯、家具使用习惯方面给我们的设计提出了很多合理化的建议。在最后进入加工阶段时，几名设计师以及家具厂里的葡萄牙工人也同样认真、执着，这对于每一件作品来说都是至关重要的。

1	2
3	4

1 Emanuel Barbosa
2 Rui Costa
3 吕诗意 Lv Shiyi
4 Paulo Providência

Portuguese friends in "Design-in-Residence Programs"

Han Feng

For us, the beauty was hided in all scenery, meal,sketch and furniture, and we could not be the absent person who take part in this amazing journey of "Design-in-Residence Programs." Fancinating stories were buit up by each fragment which we made in protugal.

There are Chinese designers in our "resident project" in this story, as well as Portuguese designers and friends.

Colorful Portuguese life

We came to Portugal with a thousand miles to participate in the "Design-in-Residence Programs" furniture design topics, it should be different from that in China. But to be different, the biggest challenge was how to find the cultural combination point of handshaking between China and Portugal in such a short period of time.

With the protuguese teachers and friends, everyone explored history and culture of this strange country in different view, such as small town like Paredes, big cities named Porto and Lisbon, the historic town named Aveiro and Guimaraens.Not only the city, but also the culture, from painting, sculpture, porcelain, clouth and furniture to their shape, patten, crafts and details.

We learned the daily routine details of town people from the morning market, experienced the local eating habits of local people from the daily meals and participated in several important holidays such as the Portuguese National Day to feel the culture and lifesyle of portugal. From the daily discussion, we try to understand the thinking of Portuguese design. Each surprise came from living details, each works had experienced a process which was self-criticism and majorization over and over again.

Portuguese designer in "Design-in-Residence Programs"

To say about the biggest features of the Portuguese designers, it must be their bushy sidewhiskers, and it bring them a design style by the sense of mature and steady. It is easy to communication and discussion with them, bucause they are easy-going.Every time we exchange our design ideas, full of exuberant expression and body language can be seen. It seemed that every words should cooperate with gesture and expression when they talk.

I extremely admired thouse portuguese designers who were addict in the work, they were not let any details go. Our discussion always continue 7 or 8 hours until midnight. They gave us some reasonable advice which were living hobbits and lifestyle in using furnitures of European. It is important to each product. Those designers and the factory works were still serious as they could also in the final process.

植物

孙滨

这次“驻场计划”使我有了和葡萄牙亲密接触的机会，体验着充满着慵懒的风情，葡式古典风格的建筑给予我强烈的视觉冲击。云淡风轻，气候宜人，让我静静地享受着这不紧不慢的日子。这里，没有宽阔的柏油马路，只有窄窄的街道和石块铺的整整齐齐的路。两侧五颜六色的房屋，装点着城市。

我们驻场的孵化器在一个偏远的小镇上。道路两侧是各有特色的小别墅，每家都精心地设计和修剪着他们的花园，美化着“脸面”。

当我第一次到达波尔图圣本托火车站时，不禁为它那高大的瓷砖壁画所震撼，其实不仅是这里，很多车站以及路上经常能够看到精美的青花瓷砖，点缀着这朴实无华的国度。

我眼中的葡萄牙是热情的，正如我所拍摄的一组植物一样，毫无掩饰地展现给所有喜爱它的人。

在这里我不禁想起：时光在懒散中滑走，摆脱夏热的纠缠，一丝凉风一丝寒，落叶悠闲、细语呢喃。

Plants

Sun Bin

This "Design - in - Residence Programs" gave me the opportunity to have close contact with Portugal and experience the idle style. Portuguese-style classical architecture gave me a substantial visual impact. The clouds are light, and the breeze blows gently to make the weather cozy. Making it possible for me to enjoy an unhurried life. There is no enough wide asphalt road. Instead, there have narrow streets and the path paved by neat stones. Colorful houses on both sides decorated the city.

Our resident incubator is in a remote town. There have unique small villas on both sides of the road. Each family designed and pruned their gardens carefully and making their homes beautiful.

I can not help but shocked by the tall ceramic tile murals when I first arrived at the Porto San Bento train station. Also, we can often see beautiful blue and white ceramic tiles on many roads and stations which decorated the unpretentious countries.

I think Portugal is passionate. And it shows to all who love it, like the set of plants I have photographed.

Here I can not help but think: time slips away in laziness, get rid of the hot summer entanglement, a trace of fresh breeze leaves leisurely and whisper.

葡萄牙街角的植物
Plants of portuguese corner

APROVEIT
10
vitaminas

记忆中大西洋东岸的色彩

闫卓远

飞机还未降落，从湛蓝的天空俯视里斯本，大片大片橙红色的屋顶在绿色的原野里汇聚，这是我对于葡萄牙的第一印象。

进入城市后我发现，如果橙红色是葡萄牙城市的俯视印象，那么橙红色天际线与各种岩石组成的墙面、伸展向远处的石块道路则组成了城市的3D效果图。橙红色主要来自于屋顶的黏土瓦，是葡萄牙传统建筑的特色，绝大部门葡萄牙建筑住宅等，都还在应用这种材料，而传统的保留，造就了葡萄牙这一有着独特魅力的景色。置身于传统建筑墙面花岗石的包围中，尤其身处古老的教堂、城堡时，仿佛一瞬间，就回到了中世纪的欧洲电影剧情中，高大的骑士举起长剑，远方传来教堂的钟声。

而点缀城市的，也在城市之外广泛分布的，就是盎然挺立的绿色了。和世界上其他国家一样，各地的城市化进程抹去了自然界中的许多碧色，依然幸运的是葡国处在一个气候宜人的地带，在葡萄牙停留的那个小镇，夏日透过层层绿叶的阳光随风影舞动，集市上和善的阿姨摆出的各种各样的盆栽，与当地人一起在夕阳下踢球的金色草坪……在回忆里停留了一幕一幕的美好。

这些色彩，因保留传统而存在，因热爱生活而存在，因向往美好而存在。到了今天，也成了我记忆中葡萄牙标志性的色彩。

Color memory of the Atlantic East Coast

Yan Zhuoyuan

The plane has not landed, overlooking the blue sky from Lisbon, a sizeable orange-red roof gathered in the green fields, this is my first impression of Portugal.

When I entered the city, I found that if the orange-red is the impression of the Portuguese city, then the orange-red skyline and walls of various rocks and the stretches of stones to the distance make up the city's 3D renderings. Orange-red clay roof tile mainly from the traditional characteristics of the Portuguese architecture, most of Portugal architectures, such as residential buildings are still using this material, and the act of traditional reservation, creating such unique and charming scene of Portugal. Surrounded by the granite walls of traditional architecture, especially in the ancient churches, castles, as if for an instant, returned to the medieval European cinema plot, a brave knight raised the sword, with the peaceful sound of a distant church bells.

And embellishment of the city, which also widely distributed outside the city, is full of vivid green. Like all other countries in the world, the process of urbanization has eclipsed many of the natural colors in the world. Fortunately, Portugal is in a climate zone with pleasant weather. The small town that I stayed in Portugal, summer sunlight dancing with the wind and shadow of cascading green leaves, the fair aunt put various potted plants, and the locals were playing in the golden grass under the sunset... These experience leaving many beautiful scenes in my memory.

These colors exist because of their traditions, their love for life, and their existence toward beauty. Today, it has also become my memory of Portugal's iconic colors.

静谧橙
Orange

盎然绿
Green

岩石灰
Grey

场景
Scene

风景
Scenery

建筑
Architecture

人物
Character

光影斑驳

赵安路

在抵达葡萄牙的第一天，我就意识到，这里的阳光比夏天的北京还要浓烈。

我们驻场的孵化器位于距离里斯本300多公里的帕雷德斯。印象中的帕雷德斯，大部分时间都被阳光笼罩，光线穿过院墙的栅栏，在地上留下斑驳的光影，这番场景给我的印象很深，也让我暂时忘却了夏日的炎热。每次出门，都会从当地居民的门前经过。偶尔还会见到在庭院中精心修整花园的房主，相视一笑。

这里的人们似乎并不在意烈日给皮肤带来的“副作用”，反而是更珍惜每一缕阳光，并享受着。倘若我在人群中撑着一把防晒伞，怕是有些格格不入，破坏了这自然景致。

即使是在首都里斯本，也鲜有像北京一样密集成群的高楼大厦。更多的是，低矮而错落有致的小楼，朴素却充满了时间所留下的印记。这里的建筑多由石块砌筑而成，相对于紫禁城中结构精妙的木质建筑，这里的教堂、城楼、石桥显得更加厚重沉稳。在阳光的照耀下，空间的层次感被凸显出来，让我沉浸其中。

行走在葡萄牙的街头，不时会有小火车从马路上经过。人们三五成群，漫步街头，互相交谈着今天的奇妙经历；有的则坐在街角的露天咖啡馆，聊天、叙旧。这些场景都被我用镜头一一记录下来，因为在这背后，还蕴藏着这座城市的文化和人们的生活态度。

我时常被这里的风景所吸引。虽都是些普通的景致，花、草、树木、道路、湖水、蓝天……但每每将这些平凡的小事物汇聚在一起，总会有不一样的惊喜。

这是一座美丽繁华的城市，为期一个月的“驻场计划”也让我感受到了不一样的风土人情。我享受这里的生活，也怀念阳光下的人们。

Light and Shadow

Zhao Anlu

On the first day of arrival in Portugal, I realized that the sun here is stronger than that in Beijing summer.

Our incubator in the factory is located in Paredes, more than 300 kilometers away from the Lisbon. Most of the time in my impression of Paredes covered by the sun, and the scene gave me a deep feeling that the light through the fences of the yard, and leaving mottled lights and shadows on the ground, which let me temporarily forget the summer heat. Every time go out, you pass by the residents, sometimes you will see a homeowner trimming the garden, then we smile at each other.

The residents seem not mind of the side effects of the scorching sun, instead, but cherish every ray of sunshine and enjoy them. If I hold a sunshade in the crowd, I'm afraid it is a bit out of tune, undermining the natural scenery.

Even in the capital Lisbon, few high-rise buildings are as dense as Beijing. There are also more scattered small buildings which are simple but full of memories of the time. Most of the buildings here made of stone masonry, and the churches, watchtowers, and stone bridges are thicker and calmer than the beautiful wooden structures in the Forbidden City. Under the sunshine, the level of space is a highlight, which allows me to immerse myself.

Walking in the streets of Portugal, from time to time there will be a small train passing through the road. People are in groups, wandering the streets, talk to each other about today's incredible experience; Some sit in an open-air cafe on the corner of the road, chatting about the old ages. I have recorded all these scenes because behind this. It also contains the city culture and people's attitude to life.

I always attracted by the scenery here. Although these flowers, grass, trees, roads, lakes, blue sky are ordinary, when you bring together these everyday little things, there will always be different surprises.

It is a beautiful and bustling city. The one-month "Design-in-Residence Programs" also made me feel different local condition and customs. I enjoy the life here and miss the people in the sun.

1 2 3
4 5 6

1 正在加工的工人
1 Workers under processing
2 工厂成品一瞥
2 Factory products glance
3 平板运输带来的思考
3 Thoughts on flat transport
4 街角一处涂鸦
4 A graffiti on the corner
5/6 家具工厂入口
5/6 Entrance of furniture factory

细节

沈相宜

记得刚抵达里斯本的那天，天空很蓝，气温很高。虽然在驱车前往帕雷德斯的路上我几度昏昏欲睡，但还是努力克服着时差所带来的困倦，试图将车窗外的异国风景尽收眼底。

我和同行其他9位设计师的起居，都被安顿在了和孵化器只有一墙之隔的居住共同体中。光是建筑外立面鲜明的现代主义设计风格就与周围的民居形成鲜明的对比。室内的家居陈设更是风格不一。后来我才了解到，这些家具都是由本土设计师完成的。

虽然这次的"设计之旅"只有短短的4周，但一切和设计相关的活动都在几位葡萄牙设计师的带领下有条不紊地进行着。有相当一部分的时间都被安排在了家具生产车间的参观。令我印象深刻的是，生产车间的一尘不染和井然有序。偶尔还会传来工人们收音机中播放的音乐声。想必在这种氛围下，工作也是一件快乐的事情吧。

尽管有时语言不通会打消我进一步沟通的念头，但这丝毫没有影响到来自葡萄牙的设计师们与我们一起进行设计方案推敲。从尺寸、结构，到材质、生产工艺，每一个细节都关系到最终设计成果的呈现。每次探讨设计方案，都要花费将近一整天的时间，也正是因为每个人对细节的严谨，才让10件产品顺利"落地"。

这次"驻场计划"让我在收获知识的同时，也得到了放松。在参观游览中我感受到了葡萄牙慢节奏的生活和每时每刻都在享受生活的人们。虽然语言不通，但也能从旁人的语气、神情中感受到他们对来自异国他乡的客人的欢迎。非常感谢这次机会带给我的难忘经历，这也是我对西方设计和文化探索的开始。

Details

Shen Xiangyi

I remember the day arrived in Lisbon.The sky is blue, and the temperature is high. Although I was drowsy on my way to Paredes, I tried my best to overcome the drowsiness caused by jet lag and try to get a panoramic view of the exotic scenery outside the window.

I was living with other nine designers and settled in a residential community with a single wall separating the incubator. The unique modernist design of the external building facade contrasts sharply with the local dwelling houses. The indoor furnishings also were a different style. Later, I knew about this furniture was done by local designers.

Although this design trip was only a short four weeks, all design-related activities were handled in an orderly fashion under the leadership of several Portuguese designers. A considerable part of the time has arranged in the furniture production workshop visit. What impressed me most was the clean and well-organized workshop. Some music could hear from the workers' radio. Presumably, in this atmosphere, working is a happy moment.

Although sometimes the language barrier would dispel my thoughts of further communication, it did not affect the designers from Portugal to work with us on the design scrutiny. From the size, structure, material, production process, every detail is related to the final design results presented. It tools almost a whole day that every time for us to discuss the design, and it is because everybody is rigorous about the details so that ten products could be "on the ground."

This "Design-in-Residence Programs" has given me a sense of relaxation as well as gaining knowledge. During the tour, I felt the slow-paced life in Portugal and the people who enjoy life at all times. Although the language barrier, but also from the tone of others, the expression of feelings they feel welcome to foreign guests. I am very grateful for the unforgettable experience this occasion brought to me. It is also the beginning of my exploration of western design and culture.

1
满园的绿植，年久的门楼，那是岁月和生命的交织

1
Garden plants, the old gate that is life and time interleaving.

2
偶然发现树洞里的蜗牛，静静的，相信它也适应了这里的慢节奏

2
The snail in the tree hole found in silence, and it believed that it also adapted to the slow pace here.

3
没有人会在意它的存在，相信它一直在这里等待，它依然那样静静的守护着

3
No one cares about it, and I believe it has been waiting here, it is still so quiet and guarded.

7
留恋于它浓烈的文化气息和艺术氛围，莱罗兄弟书店的美惊为天人

7
Love it loud cultural atmosphere and artistic atmosphere, the beauty of her brother Leiro Bookstore.

8
一片枯叶，几枚栗子，就这样装点着葡萄牙著名设计师的工作室

8
A dead leaf, a few chestnuts, and the studio of a famous Portuguese Designer.

9
破旧的路标，退了颜色却饱含了岁月的味道

9
The color of the road sign is fading but the taste of the years still here.

13
波尔图主教堂门前的雕塑，见证虔诚的教徒来来往往

13
The sculptures in front of the Porto Cathedral witness the comings and goings of devout believers.

14
波尔图火车站附近的鸽子，它们从不怕人，列队等待拍照的粉丝

14
The pigeon never afraid the human, and they wait for photo by their fans around the Porto train station.

15
这里，每一支蜡烛静静地在角落里燃烧，一份祈祷，一份牵绊

15
Here, every candle burns quietly in the corner, a prayer, a trip.

19
涂鸦是葡萄牙街景的重要元素，不乏稚拙有趣的作品

19
Graffiti is a significant element in the street of Portugal, and there have pure and exciting works.

20
波尔图大学后花园，坚硬的台阶上有很多似花非花的藓类，斑驳自然

20
Oporto university garden, there are a lot of hard steps like non flower moss, mottled nature

21
欧式古老的家具装饰，沉稳安详

21
Calm and kind shown by the decoration of ancient europ furniture.

旮旯

赵雄韬

初来葡国，这里处处充满着生机，这是个包容性很强的国家。住在帕雷德斯的小镇上，轻松，惬意，这里华人很少，路上可以感觉所有人都有着新奇的眼光：仿佛在说：快看，外国人！甚至飞驰的汽车都要慢下来回头张望，但是在这里，每个人的脸上都洋溢着快乐，充满着友善，那感觉是由心而发，不加修饰。清晨路上的清洁工、饭店服务员、超市收银员，他们很愿意和路人或顾客开个玩笑，问候几句，在葡国人的眼里貌似没有什么事比喝酒和吃饭更重要的事，他们的节奏慢得让我花了很长时间才逐渐适应。这里，没有过多的修饰，古建筑林立，我用镜头记录的是一些角落，通过这些带您走入葡萄牙的世界。这里，人们很注意自己的生活情调，不经意的一瞥，那些犄角旮旯的美，便映入眼帘。

新旧的相容随处可见，华丽不显于外表，只让人深深感受那千年的文化让我们细细品味。

Corner

Zhao Xiongtao

Early arrival in Portugal, where life is full of vitality, this is a tolerance country. Living in the town of Paredes, relaxed, comfortable, few Chinese here that can feel everyone has a new vision: as if to say: Look, foreigners! Even speeding cars slow down and look back.Here, everyone's face filled with happiness, full of friendship, that feeling without modification made from the heart. In the morning, the road cleaners, hotel attendants, supermarket cashiers, they are quite willing to talk jokes to passers-by or customers, like greetings. In the eyes of the Portuguese, nothing could be more important than drinking and eating. Their slow pace that I took a long time to adapt. Here, the ancient building everywhere without too much modification. I use the camera to record some of the corners, through which you take to the world of Portugal. Here, people pay considerable attention to their own lives, the beauty of corner come to my mind with a casual glance.

New and old compatible everywhere, gorgeous does not show in appearance, only people genuinely feel that millennial culture let us savor.

4
说它锈迹斑斑并不为过，但看起来与环境是那么的和谐，好像它本身就应该是这样的

4
It says it's not too rusty, but it looks so harmonious with the environment, as if it should be the way it is.

5
偶然抬头看见这些枯萎的枝条，像是在伪装，就好像隐藏了自己

5
A chance to look up and see the withered branches, as if they were in disguise, seemed to hide themselves.

6
不知有多少对恋人牵手走过，相拥彼此，在桥上共同许下了誓言……

6
I do not know how many lovers hand in hand, embracing each other, making a joint pledge on the bridge...

10
冰冷的铁篱挡不住热情的心，和谐至极

10
The cold iron fence can not stop the warm heart, and the harmony is extremely.

11
门牌号已然斑驳，但它记录了家的位置

11
The door number has been mottled, but it records the location of the home.

12
错综的根脉向世人展示着它的成熟，稳重

12
The root of intricate to show the world of its mature and stable.

16
饭店服务员不紧不慢的调制着各种美味，生活亦需要这样的酸甜苦辣

16
The waiter with a variety of delicious neither fast nor slow modulation, life also need such a sour, sweet, bitter, hot.

17
博物馆的内墙破旧不堪，这些陶器形状不一，完和缺的呼应

17
The interior walls of the museum are dilapidated, the shapes of these pottery are different, and the end of the museum is short.

18
凯旋门下，除了感叹雄伟外，还是感叹

18
Triumphal Arch, in addition to exclamation majestic, or exclamation.

1 2
3 4

5 6
7 8
9 10

1 建筑物外观
1 appearance of building
2 室内顶面
2 indoor top surface
3/4 地面铺装
3/4 ground pavement
5/6 帕雷德斯市政府建筑物外观
5/6 Paredes city government building appearance
7 标识
7 sign
8 葡国老师
8 Portuguese teacher
9 工厂老板与设计师
9 factory boss and designer
10 城市列车
10 city train

图案纹样

郗鑫鑫

对于此次赴葡国的设计交流，我选择了“图案纹样”的字眼作为传达葡国文化与特色的主题，从生活出发，从住居环境走起，留住那些关于帕雷德斯市的“建筑物”的艺术美。

所谓的“建筑物”主要包括两个层面，第一是建筑物的界面即墙顶地，墙面主要反映的是图案的自身纹理特色及不同材料的拼接韵律的特点。帕雷德斯市域内的院落外墙主要以毛面不规则浅色系石材拼接为主，建筑物外墙以白色石质做凸起的雕塑形象为主，当然也存在马赛克拼贴图案的境况。最有特色的是代表宗教韵味的教堂屋顶，穹顶造型做工复杂考究，客观反映了葡国特色和匠人技艺。居室顶面多利用对称手法做造型，图案精美，韵味无穷。地面铺装是葡国值得留念的东西，小碎石块的不同拼花，图案特色鲜明，材质表面肌理粗糙，韵律感较强，“直线、弧线、折线”等不同铺装手法造就了街道统一有别的传奇，第二是建筑物的眼睛，即门与窗。不同的建筑物门窗形制不同，大小不同，风格迥异，铁艺木艺石艺等多种手法交会，教堂的彩绘玻璃显示出其高贵与神圣。当然标识作为建筑物的小装饰也变得尤为重要。拍标识主要是想做标识对比，对比其文化差异的碰撞，对比其蕴涵的内在，开拓图案艺术的视野，感悟葡国文化的真谛。

一个月，只赞叹时间过得太快。回忆专业课上与葡方老师对作品的深入探讨，实践环节与工厂技师对作品的技艺研究，以及当地友人的热情激情与友情等，一幕幕就在眼前。葡国设计，匠人工艺，对待作品的认真态度深深触动着我们，希望中葡文化交流继续延续下去……

Pattern

Xi Xinxin

For this exchange of design in Portugal, I chose the word of “pattern” as the lord topic to express national culture and characteristics. Beginning with the life and the residential environment, we try to retain the “buildings” artistic beauty about Paredes.

The so-called “buildings” mainly includes two aspects. The first is that the interface buildings which constitute by ceiling, floor, and wall. The wall characteristic is primarily reflected in the texture of itself and the splicing rhythm of different materials using. In Paredes City, the courtyard wall spliced with the frosted irregular light color stone, and the exterior wall of the building mainly is raised sculpture image which made of white stone. Of course, there is also a mosaic pattern of collage. It is the lasting appeal of religion of the church dome that the workmanship is complex and exquisite reflected the characteristics and skills of Portuguese builders.The roof of the house is mostly made of plaster with symmetry, and the pattern is exquisite and the charm is infinite. It is the lasting appeal of the religion of the church dome that the craft is intricate and exquisite reflected the characteristics and skills of Portuguese builders. The roof of the house mostly made of plaster with symmetry, and the pattern is excellent, and the charm is infinite. It is the memorable things of Portugal that the pavement made of different small pieces of stone mosaic which the pattern characteristic is distinct, the material surface is roughness, and the sense of rhythm is strong.The legend of the Portuguese Pavement Street is created by the pavage techniques that are the straight line, arc, broken line and so on.

The second aspect is the eye of buildings, which means the door and window. The shape size and style of different building doors and windows are different which intersect by iron, wooden or stone art and other techniques. The stained glass of the church shows its superior and sacred. Of course, identifying as the structure of the small adornment also is particularly significant. Take a photograph for mark mainly is to want to make the comparison that comparing the cultural differences, comparing it contains internal, and develops the design art horizons, to comprehend the essence of national culture.

For a month, the time is too fast. The memory that discussed the work with the teacher of Portugal in the professional class and researched the technology with the factory technician, as well as the enthusiasm and friendship of the local friends that those were visible in front of eyes. The Portuguese builders' design process, serious attitude towards work deeply touched us, hope that the Chain and Portuguese cultural exchanges continue.

INTRODUCTION

简介

杨琳，1968年生。副教授，任教于北京建筑大学建筑与城市规划学院设计学系，目前主要在建筑学、设计学学科，以及环境设计专业领域从事教学、科研和工程设计工作。近年主要在传统文化与环境设计、商业环境设计等方向进行设计及研究，发表“模数协调标准在室内设计中的应用”“BIM技术在室内设计中的应用”“明式家具数字化构法与技艺保护探讨”等论文，主持《室内细部设计资料集》的分册主编工作；锦州建国酒店设计荣获2013年第十届精瑞科技奖室内设计金奖。

Yang Lin, born in 1968, is an associate professor who taught in the School of Architecture and Urban Planning, Beijing University of Civil Engineering and Architecture. Currently, he is mainly engaged in teaching, research, and engineering in the fields of architecture, design, and environment design. In recent years, he has primarily designed and researched on the aspects of traditional culture and environment design and business environment design. He published "Application of Modular Coordination Standards in Interior Design," "Application of BIM Technology in Interior Design," "Ming-Style Furniture Digital Structure and Skills Protection "and other essays. And he presided over the book " Interior Design Information Collection." The design of Jinzhou Jianguo Hotel won the 2013 Jing Rui Technology Award Interior Design Gold Award.

邹乐，1992年生，北京市人。2010~2014年，北京建筑大学工业设计专业毕业，获工学学士学位。2014~2017年，北京建筑大学设计学专业毕业，室内设计研究方向，获艺术学硕士学位。在校期间参与中葡第二期“驻场计划”，设计作品在2016北京国际设计周展出；参与“苏州造物”明式家具（苏作）制作技艺精品展、商丘文化艺术中心等相关项目的设计工作。目前在北京市建筑设计研究院从事室内设计工作，参与的设计项目包括腾讯北京总部、联想北京总部、小米北京总部、京东北京总部、北京新机场、中石化科研楼等。

Zou Le, born in Beijing city in 1992, From 2010 to 2014, graduated from Beijing University of Civil Engineering and Architecture with a bachelor's degree in industrial design. From 2014 to 2017, Beijing University of Civil Engineering and Architecture graduated with a major in Design and got a master's degree in Art and Design in a space environment and facilities design.During the period of school, participated in the "Phase II of Design-in-Residence Programs" and the design work exhibited in the Beijing Design Week 2016; participated in the design of "Suzhou creation" Ming-style furniture (Su Zuo) production art exhibition, Shangqiu Cultural Arts Center and other related projects. At present, engaged in interior design in Beijing Institute of Architecture Design, and take part in the design of Tencent Beijing headquarters, Lenovo Beijing headquarters, Xiaomi Beijing headquarters, Jingdong Beijing headquarters, Beijing new airport and Sinopec Research Building.

朱宁克，1977年生。北京理工大学设计艺术学硕士学位，任教于北京建筑大学建筑与城市规划学院设计学系，中国建筑学会室内设计分会会员，中国建筑装饰协会会员，北京工业设计促进会会员。主要从事环境设计、工业设计、艺术设计、数字化设计等工作。出版有《Autodesk Revit architecture 2010建筑设计速成》《数字化建筑设计概论》《界面视觉传达》等著作，发表论文《Using BIM Technology to Optimize The Traditional Interior Design Work Mode》《产品设计与建筑设计在空间形态中的互融》《构筑建筑数字化设计教学平台》等数篇；完成设计项目“锦州大酒店室内设计”“内蒙古巴音淖尔医院室内设计”等，指导学生获得奖项数十项。

Zhu Ning Ke, born in 1977, MFA of Beijing Institute of Technology. Currently, he taught in the school of architecture and urban planning, Beijing University of Civil Engineering and Architecture, member of China Institute of Interior Design, member of China Building Decoration Association, member of Beijing Industrial Design Promotion Organization. Mainly engaged in environment design, industrial design, art design, digital design. And he published crash course of architecture design by Autodesk Revit architecture, Generality of Digital Architecture Design, Visual communication of interface. And he issued theses which are Using BIM Technology to Optimize The Traditional Interior Design Work Mode, Integration of Protect Design and Architecture Design in Space Form, Establishment of Digital Design Teaching Platform, etc.. Complete the design project of the interior design of Jinzhou hotel, the interior design of Inner Mongolia Bayan Nur hospital. And guide students to get many kinds of awards.

韩风，1977年生。毕业于清华大学美术学院环境设计专业，获艺术学博士学位，于美国威斯康辛大学人文生态学院室内设计系获博士学位。现任教于北京建筑大学建筑与城市规划学院设计学系，北京清尚建筑设计研究院第十一工作室任设计经理。长期从事建筑、景观、室内及中国传统文化等领域的教学与设计实践工作。留美期间，在威斯康辛州国际交流中心、威斯康辛州红堡艺术馆举办多场建筑绘画作品个人展览。

研究领域集中在生态审美、可持续设计理论与实践，出版了多部专著，在多篇核心期刊论文成果，荣获“为中国而设计”“艺术与科学国际设计展”等多项设计大奖，北京泰富酒店荣获2017年中国设计博览会金奖。其代表作品涉及规划、景观、建筑及室内等领域，包括：北京泰富酒店（五星级）、北汽采育国际会议中心（四星级）、洛阳天堂博物馆、北京灵璧石博物馆、山西剪纸艺术博物馆、内蒙古达茂旗文化景观雕塑、北京中福丽宫科创园、北京中林置业大厦等项目。

Han Feng, born in 1977, received his Ph.D. degree in art from environment design major of Academy of Arts & Design, Tsinghua University, and received a doctoral degree from interior design department of College of Humanities and ecology, University of Wisconsin. Currently, he taught in the school of architecture and urban planning, Beijing University of Civil Engineering and Architecture, and assume the office of the design manager of the eleventh studio of the Beijing Tsingshang Architecture Design and Research Institute. He mainly engaged in the teaching and design practice in the fields of architecture, landscape, interior and Chinese traditional culture for a long time. During the period of stay in the United States, some individual exhibitions of architectural paintings held at the International Center of Wisconsin and the Red Gym Art Center of Wisconsin.

His research focused on ecological aesthetic, sustainable design theory and practice. He published several monographs and research achievements in various core journals, won some design awards such as "Design for China," "Arts and Science International Design Exhibition," and his design of Beijing Tylfull Hotel won the gold award in Chinese Design Exhibition in 2017, etc. The representative work involved in planning, landscape, architecture and interior areas. It included Beijing Tylfull Hotel (five-star), BAIC Caiyu International Conference Center (four-star), Luoyang Tiantang Museum, Beijing Lingbishi Museum, Shanxi Paper-cutting Arts Museum, Cultural landscape sculpture of Inner Mongolia Darhan Muminggan Joint Banner, Beijing Zhongfuligong International Science and Technology Park, Beijing Zhonglin Property Building.

孙滨，1983年生，小学高级教师。毕业于首都师范大学初等教育学院，现任朝阳区实验小学艺术教育主管、朝阳区区级骨干教师、北京市教育学会“开发大脑潜能，发展形象思维”研究会会员。主要研究方向为美术教学研究。在北京市多项论文比赛中取得奖项，曾获得北京市首届“智慧教师”征文一等奖，北京市“京美杯”一等奖，参与中葡第二期“驻场计划”，设计作品在2016北京国际设计周展出。

Sun Bin, born in 1983, a senior teacher of primary school, graduated from Elementary Education College, Capital Normal University, At present, as an Art Education Director of Beijing Chaoyang Experimental Primary School, Chaoyang District Backbone Teacher, and the member of research and development of image thinking of Beijing Institute of Education. The primary research direction is fine arts teaching research and has won awards in many papers competition in Beijing. Won the first prize in the first Beijing "smart teacher" essay award and the first prize of Beijing "Jingmei Cup." Participated in the "Phase II of Design-in-Residence Programs" and the design work exhibited in the Beijing Design Week 2016.

闫卓远，1992年生，河南省漯河市人。2010~2014年，沈阳工业大学工业设计专业毕业，获工学学士学位。2014年~2017年，北京建筑大学就设计学专业毕业，空间环境与设施产品设计研究方向，获艺术学硕士学位，被评为2017届北京市优秀毕业研究生。

在校期间参与了中葡第二期“驻场计划”，设计作品在2016北京国际设计周展出;参与“苏州造物”明式家具(苏作)制作技艺精品展、“中葡设计握手·家具篇”展览的设计工作；参与故宫博物院养心殿保护性研究“瓦作”调研项目。目前在中国科学技术馆从事展品形式设计工作。

Yan Zhuoyuan, born in Luohe City, Henan Province in 1992. From 2010 to 2014, graduated from Shenyang University of Technology with a bachelor's degree in industrial design. From 2014 to 2017, Beijing University of Civil Engineering and Architecture graduated with a major in Design and got a master's degree in Art and Design in the space environment and facilities design. He honored as one of the outstanding graduates in 2017 in Beijing.

During the period of school, participated in the "Phase II of Design-in-Residence Programs" and the design work exhibited in the 2016 Beijing Design Week. Participated in the design of "Suzhou creation" Ming-style furniture (Su Zuo) production art exhibition and"Handshake between Chinese and Portuguese Design. Furniture" exhibition. Participate in the study of protective tile repair research in Hall of Mental Cultivation. Currently, work in the China Science and Technology Museum and engaged in the design of exhibition display.

赵安路，1994年生。毕业于北京建筑大学工业设计专业，获工学学士学位。现从事于交互设计、平面设计、用户体验设计等。在校期间参与中葡第二期“驻场计划”，设计作品在2016北京国际设计周展出。

Zhao Anlu, born in 1994. She Graduated from Beijing University of Civil Engineering and Architecture, major in Industrial Design, Bachelor's degree in Engineering. During the period of school, participated in the "Phase II of Design-in-Residence Programs" and the design work exhibited in the Beijing Design Week 2016.

沈相宜，1993年生，北京市人。毕业于北京建筑大学工业设计专业，获工学学士学位。现就读于不来梅艺术学院硕士研究生，专业方向为整合设计。在校期间参与中葡第二期“驻场计划”，设计作品在2016北京国际设计周展出。

Shen Xiangyi, born in Beijing 1993, graduated from Beijing University of Civil Engineering and Architecture, major in Industrial Design, Bachelor's degree in Engineering. Now studying at University of the Arts Bremen, major in Integrated Design. During the period of school, participated in the "Phase II of Design-in-Residence Programs" and the design work exhibited in the 2016 Beijing Design Week.

赵雄韬，1983年生，北京市人，小学高级教师，北京市美术骨干教师。毕业于首都师范大学初等教育学院美术专业，获教育学学士学位。现任北京市朝阳区实验小学美术教师。近年主要研究少儿版画，版画作品入选2016年第四届广州国际藏书票暨小版画双年展，摄影作品曾获“竞园杯”摄影大赛铜质收藏奖，撰写文章《可以玩的版画》发表在国家艺术类核心刊物《中国中小学美术》，参与中葡第二期“驻场计划”，设计作品在2016北京国际设计周展出。爱好摄影、滑雪。

Zhao Xiongtao, born in Beijing in 1983, primary school senior teacher, Beijing fine arts Backbone Teacher. Graduated from Elementary Education College, Capital Normal University, and major in fine arts. Currently, teaching in Beijing Chaoyang Experimental Primary School. The primary research is children's prints, and his print was selected in the fourth Biennial exhibition of Guangzhou international Exlibris in 2016 and won the bronze collection award in "Jingyuan Cup" photography competition. The essay "prints could play" published in national art journals "Chinese Primary and Secondary School Art." Participated in the "Phase II of Design-in-Residence Programs" and the design work exhibited in the Beijing Design Week 2016. Fond of photography and skiing.

郗鑫鑫，1988年生，山东省莱芜市人。北京建筑大学艺术学硕士学位，中国室内装饰协会会员。师从高丕基教授、梁思佳总工程师，研究方向重点在公共环境艺术设计、室内设计、住宅装修部件化设计研究等。

2017年度参与齐鲁制药产业园景观规划设计、山西太原某酒店大堂方案设计；

2016年度担任玛格定制家具设计总监，家具作品《小木马》参展北京国际设计周；

2015年度于中国建筑设计研究院有限公司参与北京大兴某办公楼建筑及室内设计；

2014年度参与辽阳市矿山生态修复工程湿地区域设计、细河景观带设计。

2013年度于北京冠亚伟业民用建筑设计有限公司参与银川美术馆室内精细化设计、参与沈阳银基售楼处室内设计、参与河南烟厂室内精细化设计方案、参与洋河蓝色经典酒厂室内设计、清华大学建筑学院古村落研究所罗德胤组、参与河南省新县西河村保护与修复现场调研、参与河南省新县丁李湾村调研。

Xi Xinxin, born in Laiwu City, Shandong Province in 1988, MFA of Beijing University of Civil Engineering and Architecture, member of China National Interior Decoration Association. A study from Professor Gao Pi Ji and Liang Sijia chief engineer. Research focuses on the public environment art design, interior design, and residential decoration design.

In 2017, participated in the landscape planning and design of Qilu Pharmaceutical Industrial Park and the design of a hotel lobby in Taiyuan, Shanxi;

In2016, as Mage Custom Furniture Design Director, and furniture work "Little horse" display on Beijing Design Week;

In 2015, participated in the construction and interior design of office building in Beijing Daxing with China Architecture Design Group;

In 2014, participated in the design of the wetland ecological rehabilitation project and Xihe river landscape belt of the mine restoration project in Liaoyang City.

In 2013, participated in the design of the interior of Yinchuan Art Museum with Beijing Guanyaweiye Civil Architectural Design Co., Ltd, the interior design of Shenyang Yinji Sales Office, the plan design of Henan Tobacco Factory, and the interior design of Yanghe Blue Classic Winery. Participated in Ruo Yinde group, Institute of ancient villages, School of Architecture, Tsinghua University, and involved in the study of protection and recovery in Hexi village, Xin County, Henan Province and Liwang Village, Xin County, Henan Province.

INDEX

索引

Ming-style Furniture 明式家具

四出头官帽椅
Armchair with four Protruding ends
589 x 493 x 1127 (mm)

P 016

灯挂椅
Lamp-hanger chair
490 x 428 x 1090 (mm)

P 022

圈椅
Round-backed armchair
704 x 605 x 1022 (mm)

P 028

六方形南官帽椅
Southern official's hat armchair
800 x 563 x 962 (mm)

P 034

玫瑰椅
Rose chair
612 x 460 x 828 (mm)

P 040

一腿三牙罗锅枨方桌
Three spandrels humpbacked strecher Squared table
980 x 980 x 880 (mm)

P 046

夹头榫带屉板平头案
Elongated bridle joint flat-top recessed-leg table with drawers
707 x 371 x 790 (mm)

P 052

翘头案
Head desk
1400 x 463 x 836 (mm)

P 058

ADA

Modern Furniture　现代家具

总统桌
President table
2500 x 800 x 785 (mm)

P 067

总统椅
President chair
750 x 750 x 1070 (mm)

P 073

咖啡几
Coffee table
∅ 1050 x 492 (mm)

P 079

缠绕
Binding
∅ 650 x 830 (mm)

P 085

妆・纳
Sensation
547 x 290 x 820 (mm)

P 091

分・格空间
Separate space
规格Dimension
600 x 300 x 600 (mm)
(正方形单元组合)
(Square Unit)

P 097

清风
Brezze
906 x 320 x 1115 (mm)
300 x 258 x 300 (mm)

P 103

一方凳
A cube
263 x 263 x 353 (mm)

P 109

融错・空间
The interlaced dimensions
442 x 447 x 1998 (mm)

P 115

小木马
The little horse
404 x 725 x 498 (mm)

P 121

后记

《中葡设计握手　家具篇》一书即将付梓。从2015年6月28日—7月23日，北京建筑大学等院校参加北京市人民政府外事办公室组织的第二期“中国青年设计师驻场四季计划”（以下简称“驻场计划”）开始，继之2016年9月9日在2016北京国际设计周活动中举办“中葡设计握手·家具篇——从北京到帕雷德斯”展览暨中外交流会，再至“艺勇军”学术团队一年多的辛勤收集、研究、编制成卷，历经三载，本书受到校内外各方的关心、帮助和支持。

以1907年清代京师初等工业学堂为办学开端，砥砺发展而来的北京建筑大学，给了我们开展设计艺术领域研究的学科环境和人才培养的专业环境，借此记载传扬。

在学科建设上，本校设计艺术领域学科建设始于1980年建筑学学科初创时期的室内设计及其理论方向和美术学方向；2007年，获得设计艺术学硕士学位授权二级学科点，学校增列文学学科门类；2012年，设计艺术学对应调整为设计学一级学科，学校对应调整增列了艺术学学科门类；2014年，增列工业设计工程硕士专业学位授权领域点。

在专业建设上，本校于1980年增列建筑学专业时同步设置了室内设计方向，1991级～1992级实行专业方向分流，设置建筑学专业（五年制）室内设计方向班；2000年增列工业设计专业，突出建筑环境与设施产品设计方向特色；2010年增列环境设计专业（2013年起开始招生），构建了艺术学门类本科教育基础。至此，本校完成工学门类下“工业设计—工业设计工程”、艺术学门类下“环境设计—设计学”的“本硕一体”学科专业系统和设计艺术领域人才培养平台构建。

2012年，本校获批服务国家特殊需求“建筑遗产保护理论与技术”博士人才培养项目；2014年，获批设立“建筑学”博士后科研流动站；2015年10月，北京市人民政府与住房和城乡建设部签署共建协议，本校正式进入省部共建高校行列；2016年5月，本校“未来城市设计高精尖创新中心”获批“北京高等学校高精尖创新中心”；2018年1月，学校成立“设计艺术研究院”，通过整合创新，开启了进一步加强设计学学科建设，进一步凝练学科方向特色，加大对设计艺术专业人才培养的支撑力度，发挥学科专业优势全面服务北京“四个中心”特别是文化中心建设发展的新阶段。2018年5月，本校获批博士学位授予单位，建筑学、土木工程获批博士学位授权一级学科点，北京建筑大学正式迈入博士学位授予单位新时代！踏上努力建设国内一流、国际知名、具有鲜明建筑特色的高水平、开放式、创新型大学的新征程。

感谢北京市人民政府外事办公室、中国驻葡萄牙大使馆、葡萄牙驻华大使馆、北京市国有文化资产监督管理办公室、文化和旅游部恭王府博物馆、北京工业设计促进中心、北京歌华文化发展集团、葡中文化协会等在“驻场计划”、北京国际设计周等活动中对北京建筑大学的指导和帮助！感谢中外相关媒体积极地宣传报道！

感谢北京建筑大学对设计艺术研究院和设计学学科建设工作的指导和支持！

感谢张爱林校长百忙之中为本书作序！

感谢著名木刻篆刻艺术家傅稼生先生为本书刊石！傅先生治印书法成为以书籍设计诠释中葡家具设计文化和创意交流成果的点睛之笔。

感谢校内外各级领导、行业专家、同事、朋友们的关心、支持和帮助！

感谢中国建筑工业出版社的大力支持和帮助！

“吾家洗砚池头树，个个开花淡墨痕。不要人夸好颜色，只留清气满乾坤。”（引自[元]王冕《墨梅》）。借本书出版，抛砖引玉，以促进我们的研究工作做得更好，期望将设计艺术领域的合作与交流之手握得更紧。

二〇一八年五月二十一日（戊戌・小满）

北京建筑大学设计艺术研究院
传统家具构法与制作技艺保护研究所
金点“创・空间” 艺勇哲作

Postscript

"Handshake between Chinese and Portuguese design. Furniture - From Beijing to Paredes" will be published. Started from some university participated in "Phase | | Programs of Chinese Young Designers Design in Residence in Four Seasons"(Design-in-Residence Programs) which was organized by Foreign Affairs Office of the People's Government of Beijing Municipality, between June 28th and July 23rd, 2015. Then the exhibition "furniture. Handshake between Chinese and Portuguese design -- from Beijing to Paredes" & Chinese and foreign exchanges succeed on September 9th, 2016. Finally, collecting, researching and compiling by YiYongJun Academic Group for more than a year. It received concern, help, and support from the school or outside school community during this three years.

Beijing University of Civil Engineering and Architecture which was the Imperial Primary industry school that established in Qing dynasty in 1907, give us a professional environment of the subject and personnel training in the field of Design Arts research that should be recorded and spread.

With discipline development, Design arts subject of BUCEA was start from interior design and its theory and fine art in the Start-up period of architecture in 1980.Master of Design Arts was authorized to the second discipline, and it adds literature as a university discipline category in the year of 2007. Design Arts discipline adjusted to the first level discipline, and it added art as a university discipline category in 2012. It added degree authorization of Industrial Design Engineering professional degree in 2014.

With the expansion of major, BUCEA set the major of interior design and architecture at the same time in 1980. It implemented major diversion with Grade 1991 and 1992 and set the interior design class of architecture (five - year education).It added industrial design as a major which highlight feature of building environment and facility Product Design in 2000.It added major of environment design, start enrollment from 2013, which established a foundation of undergraduate education of arts. So far, BUCEA finished construction of bachelors-masters integration and personnel training platform of design and arts that is "Industrial Design - Industrial Design Engineering" in engineering category and "Environment Design - Design" in the arts category.

BUCEA had already approved the doctor training program named Architectural Heritage Protection Theory and Technology for a special national request in 2012. It Set up Postdoctoral research station of architecture in 2014. BUCEA into the university list which is co-sponsored by province and ministry that sign by The People's Government of Beijing Municipality and Ministry of Housing and Urban-Rural Development in October 2015. Beijing Advanced Innovation Center for Future Urban Design enter into University of Beijing Advanced Innovation Center in May 2016. BUCEA established Institute of design and art for integration and innovation, in January 2018. It strengthened the discipline construction and coagulated particular direction of design science. It steps up supports to personnel training in design and art, and it gives play to the professional advantage in servicing Beijing "Four Centers," especially the new stage of development of the cultural center.

Thanks for guide and help for BUCEA from Foreign Affairs Office of the People's Government of Beijing Municipality, Chinese Embassy in Portugal, Portuguese Embassy in China, State-owned Cultural Assets Supervision and Administration Office of the People's Government of Beijing Municipality, The Museum of Prince Kung's Palace, Beijing Industrial Design Center, Beijing Gehua Cultural Development Group and ACLC - Associacao Cultural Luso Chinesa in Design-in-Residence Programs and activities of Beijing Design Week! Thanks for vigorous propaganda from Chinese and foreign media!

Thanks for support and guidance of discipline construction from Academic of design and arts of Beijing University of Civil Engineering and Architecture.

Thanks headmaster Zhang Ailin for writing the preface sincerely!

Thanks for carving work which is the essence of book design from illustrious carving artist Fu Jiasheng! It interprets the furniture design culture between Chinese and Portuguese and the achievement of creative communication!

Thanks for the concern, support, and help from all leaders, industry experts, colleagues, and friends!

Thanks for sturdy support and help from China Building Industry Press!

"The long on my washing inkstone pool of plum trees, flowers have pale ink marks. Not angling for compliments, I'd be content that my integrity fills the universe, "from Wang Mian, Ink plum. Hoping we can do better in research work and we expect the cooperation and communication handshake will work more closely through the publication of this book.

2018.05.21(WuXu Year · Grain buds)

Academic of design and arts of Beijing University of Civil Engineering and Architecture
Institute of Traditional Furniture Structure Methodologies and Protection Techniques Research
Gold Point "Innovation Space" Yi Yong Zhe Zou Studio